AF845308

Evaluation of a Real-Time Online Monitoring Method for *Cryptosporidium*

The mission of the Awwa Research Foundation (AwwaRF) is to advance the science of water to improve the quality of life. Funded primarily through annual subscription payments from over 1,000 utilities, consulting firms, and manufacturers in North America and abroad, AwwaRF sponsors research on all aspects of drinking water, including supply and resources, treatment, monitoring and analysis, distribution, management, and health effects.

From its headquarters in Denver, Colorado, the AwwaRF staff directs and supports the efforts of over 800 volunteers, who are the heart of the research program. These volunteers, serving on various boards and committees, use their expertise to select and monitor research studies to benefit the entire drinking water community.

Research findings are disseminated through a number of technology transfer activities, including research reports, conferences, videotape summaries, and periodicals.

Evaluation of a Real-Time Online Monitoring Method for *Cryptosporidium*

Prepared by:
Gregory M. Quist
The LXT Group, 325 Rock Ridge Place, Escondido, CA 92027
(formerly at PointSource Technologies)

Ricardo DeLeon
The Metropolitan Water District of Southern California
P.O. Box 51453, Los Angeles, CA 90054-0153

and

Ira Cecil Felkner
9033 Bay Harbour Circle, West Palm Beach, FL 33411
(formerly at PointSource Technologies)

Sponsored by:
Awwa Research Foundation
6666 West Quincy Avenue, Denver, CO 80235-3098

Published by:

Awwa Research Foundation
Advancing the Science of Water®

DISCLAIMER

This study was funded by the Awwa Research Foundation (AwwaRF). AwwaRF assumes no responsibility for the content of the research study reported in this publication or for the opinions or statements of fact expressed in the report. The mention of trade names for commercial products does not represent or imply the approval or endorsement of AwwaRF. This report is presented solely for informational purposes.

Copyright © 2004
by Awwa Research Foundation

All Rights Reserved

CONTENTS

TABLES

FIGURES

FOREWORD

The Awwa Research Foundation is a nonprofit corporation that is dedicated to the implementation of a research effort to help utilities respond to regulatory requirements and traditional high-priority concerns of the industry. The research agenda is developed through a process of consultation with subscribers and drinking water professionals. Under the umbrella of a Strategic Research Plan, the Research Advisory Council prioritizes the suggested projects based upon current and future needs, applicability, and past work; the recommendations are forwarded to the Board of Trustees for final selection. The foundation also sponsors research projects through an unsolicited proposal process; the Collaborative Research, Research Applications, and Tailored Collaboration programs; and various joint research efforts with organizations such as the U.S. Environmental Protection Agency, the U.S. Bureau of Reclamation, and the Association of California Water Agencies.

This publication is a result of one of these sponsored studies, and it is hoped that its findings will be applied in communities throughout the world. The following report serves not only as a means of communicating the results of the water industry's centralized research program but also as a tool to enlist the further support of the nonmember utilities and individuals.

Projects are managed closely from their inception to the final report by the foundation's staff and large cadre of volunteers who willingly contribute their time and expertise. The foundation serves a planning and management function and awards contracts to other institutions such as water utilities, universities, and engineering firms. The funding for this research effort comes primarily from the Subscription Program, through which water utilities subscribe to the research program and make an annual payment proportionate to the volume of water they deliver and consultants and manufacturers subscribe based on their annual billings. The program offers a cost-effective and fair method for funding research in the public interest.

A broad spectrum of water supply issues is addressed by the foundation's research agenda: resources, treatment and operations, distribution and storage, water quality and analysis, toxicology, economics, and management. The ultimate purpose of the coordinated effort is to assist water suppliers to provide the highest possible quality of water economically and reliably. The true benefits are realized when the results are implemented at the utility level. The foundation's trustees are pleased to offer this publication as a contribution toward that end.

Walter J. Bishop
Chair, Board of Trustees
Awwa Research Foundation

James F. Manwaring, P.E.
Executive Director
Awwa Research Foundation

ACKNOWLEDGMENTS

The authors wish to thank the following organizations, utilities and individuals for their help and support in this project:

- Awwa Research Foundation (AwwaRF) for the support and nurturing of this project
- AwwaRF project managers—Kathryn Martin, Jarka Popovicova, and Misha Hasan—for providing guidance and help on this project
- The Project Advisory Committee members who provided helpful and insightful comments: Gary Burlingame (City of Philadelphia Water Department, Philadelphia, PA), Ron Fayer (U.S. Department of Agriculture, Beltsville, MD), and Annette Prescott (Thames Water, London, England)
- The water utilities and individuals at the following utilities that provided many water samples and staff time for sample collection:
  - The Metropolitan Water District of Southern California
  - Orange County Water District
  - Irvine Ranch Water District
  - The City of Escondido
  - Rincon del Diablo Metropolitan Water District

In addition, the authors wish to thank the following individuals for their help on this project: Anagha Chitre, Daffodil Fallar, Mark Williams, and Connie Chou at The Metropolitan Water District of Southern California (MWD), Los Angeles, CA; and David Drake, Trung Lam, Michelle Charron, David Hunsberger, Hanno Ix, Nick Iorio, Maria Tikkanen and Alex Aguirre at PointSource Technologies (PointSource), Escondido, CA; and Bill Fitzgerald (Cambridge University) for providing technical assistance in the areas of laboratory measurements, mathematical analysis, and computer code development. In addition, the authors are indebted to Paul Rochelle (MWD) for providing technical guidance on analysis of *Cryptosporidium*; Lynn Kelemen (MWD) and Gil Granito (PointSource) for assistance with contracts and administrative aspects; and Helene Baribeau of Carollo Engineering, Santa Ana, CA, for supporting the initial development of the proposal while at MWD.

The authors wish to thank Jen Clancy of Clancy Environmental Consultants, Inc. (St. Albans, VT), Becky Hoffman of Wisconsin State Laboratory of Hygiene, and Marilyn Marshall of Sterling Parasitology Lab of the University of Arizona for providing timely and high-quality *Cryptosporidium* samples and technical guidance.

The authors wish to thank Philip Wyatt for his incredible vision, insight, and work over the past 40 years which has made it possible for real-time monitoring to become viable and useful for the water industry.

Finally, the authors are deeply appreciative of Melba Means and Helga Freitag (PointSource) for providing administrative and moral support.

EXECUTIVE SUMMARY

Rapid, reliable and accurate detection of microorganisms in drinking water is a longstanding and important goal for the drinking water industry. *Cryptosporidium parvum* is of particular concern because it exhibits high resistance to disinfectants at the doses applied in water treatment plants, has low infectious dose, and no drug approved for prophylaxis or therapy is currently available. Current on-line monitoring methods such as turbidimeters and particle counters are commonly used as indicators to warn of potential contamination problems based on changes in turbidity or particle size distributions. However, these techniques are inadequate for providing accurate or reliable monitoring for the presence of microorganisms such as *Cryptosporidium*. (Nieminski et al., 2000) The Las Vegas *Cryptosporidium* incident in 1994 (Goldstein et al., 1996) is an example of an outbreak associated with a filtered water supply apparently meeting all appropriate standards for that time. Current microorganism monitoring methods rely upon sporadic grab samples and delayed laboratory results. These methods provide no effective health benefit because by the time results are known, the water has already passed into the distribution system and likely consumed.

The purpose of this project was to evaluate and test a promising technology known as multi-angle light scattering (MALS), that has the potential to provide real-time, on-line, continuous monitoring for pathogens in drinking water.

MALS operates by detecting scattered light from individual particles as they pass through a laser beam. The light scattering pattern generated by a particle is determined by the particle's size, refractive index, shape, internal structure, and surface features (c.f. Wyatt, 1975). Differences in features that are much smaller than the wavelength (in the present case, the wavelength is 685 nm) such as DNA are not resolvable, so this technique cannot detect differences in organisms at the genetic level. MALS classifies particles that pass through the beam by comparing the scattered light pattern with pre-existing "libraries" of scattered light patterns from organisms and particles of interest including, for example, typical matrix particles commonly found in treated water (c.f. Quist and Ix, 2003).

The viability and utility of the MALS technique for use in by drinking water industry are a function of its ability to distinguish microorganisms of interest, such as *Cryptosporidium parvum*, from harmless background matrix particles under real water conditions.

RESEARCH OBJECTIVES

The Awwa Research Foundation funded this study to provide a proof-of-concept of the MALS technique to evaluate the sensitivity and accuracy under various background matrix water conditions. *Cryptosporidium parvum* was chosen as the initial target organism for testing for several reasons. First, *Cryptosporidium* is difficult to detect and enumerate using state-of-the-art methods such as EPA Method 1623. Second, the 1993 Milwaukee incident (Mackenzie et al., 1994) highlighted *Cryptosporidium* as an organism of high interest to the water industry, particularly because *Cryptosporidium* resists standard treatment techniques. Lastly, in its environmental stage, *Cryptosporidium* forms relatively stable oocysts that are relatively insensitive to variable environmental conditions such as temperature and nutrients in the matrix water.

In addition, the MALS technique was tested to determine if it could provide sufficient sensitivity to detect morphological changes (if any) of *C. parvum* oocysts caused by various treatments or excystation. Success of the ability to detect morphological changes due to treatment would be an indication of the broader applicability of MALS to detect other waterborne organisms that have distinct morphologies from *Cryptosporidium* oocysts.

The specific aims of the research were twofold:

1. To evaluate the MALS technique for detecting and enumerating *Cryptosporidium* in various water types, ranging from purified reagent-grade water to typical source and treated surface and ground waters.
2. To evaluate the ability of the MALS technique to discriminate between viable untreated oocysts and those that are physiologically impaired through treatment (ozone, heat, and ultraviolet – "UV") or excystation.

APPROACH

The MALS system was tested using various water types, including purified laboratory water, finished drinking water, ground water, and untreated surface water. This investigation led to the discoveries that: (a) MALS performance appears to be a function of particle content – concentration and size distribution - not turbidity as originally assumed; and (b) *Cryptosporidium* oocysts were not properly accounted for when seeded into real (not highly purified) drinking waters. Based on these results, the original project scope was modified to determine where oocyst losses were taking place and to determine if those losses could be minimized. A series of tests were performed to evaluate the potential reasons for the unaccounted *Cryptosporidium* oocysts, including repeated experiments, spiking concentration testing, and changes in the methodology used to train the MALS system to recognize oocysts in real waters as opposed to purified laboratory water. In addition, several modifications were made on the MALS measurement system, including changes in the opto-electronics, analysis software and signal processing. The effects of these changes on the ability to detect oocysts were measured.

Finally, a series of tests using untreated viable oocysts compared to oocysts undergoing heat treatment, ozone disinfection, UV inactivation, and excystation were performed in order to determine if MALS could distinguish between untreated and treated/excysted oocysts. The level of inactivation, as determined by cell culture infectivity, was measured in conjunction with the MALS measurements.

MAJOR FINDINGS

Major findings from the proof-of-concept effort were as follows:

1. It was possible to distinguish *Cryptosporidium parvum* oocysts from matrix water particles in a controlled environment with an example of typical finished drinking water (effluent from the Metropolitan's Water District of Southern California's Weymouth Filtration Plant, in La Verne California). This was done in real-time, continuously using an on-line MALS system in a laboratory setting. Identification Rates (the ratio of *Cryptosporidium* oocysts identified correctly to the total number of oocysts detected passing through the system) of 11% to 45% were achieved, with accompanying False

Positive Rates (the ratio of matrix water particles incorrectly identified as *Cryptosporidium* to the total number of matrix particles passing through the system) of 0.3% to 3%.

2. There is a trade-off between Identification Rate and False Positive Rate. This trade-off is expected and typical of other detection techniques, such as in radar applications. A higher Identification Rate can be achieved at the cost of a higher False Positive Rate, and conversely a lower False Positive Rate can be achieved at the cost of a lower Identification Rate. The MALS system can be "tuned" via software to provide either case, or something in the middle, for a potential user of the system, based on needs and requirements.
3. The ability to detect *Cryptosporidium* oocysts in finished drinking water by MALS is a function of the quality of system electronics and detection software. Initially, it was found that while *Cryptosporidium* could be easily identified in purified laboratory water, it was difficult to identify in finished drinking water. Once the reasons for this – electronics and signal processing – were discovered and corrected, detection rates significantly improved. This is analogous to using an unfocused microscope to see microorganisms in a sample – focusing increases fidelity and accuracy.
4. The MALS system was capable of discriminating live oocysts from those treated by heat and ozone. The system can discriminate excysted from non-excysted oocysts. It is apparent from these results that heat treatment and ozone treatment generate morphological changes significant enough in the oocysts to be detectable by MALS. In the case of heat treatment, the internal sporozoites are damaged or destroyed, and in the case of ozone treatment, there is significant surface and morphological damage to the oocysts.
5. The MALS system could not discriminate untreated from ultraviolet (UV) light treated oocysts, in these experiments. This is not an unexpected result. Sufficient doses of UV light damage the DNA in the sporozoites and this small-size scale damage is not detectable by light scattering using wavelengths that are much larger than the damage.
6. The Estimated Limit of Detection (ELOD) was determined for both purified laboratory water and a sample of finished drinking water. Since the MALS technique is a continuous real-time technique, the ELOD is a function of time – longer time results in more water patrolled by the MALS system. For purified laboratory water, the ELOD for *Cryptosporidium* oocysts was found to be 7 oocysts/mL, 0.7 oocysts/mL, and 0.1 oocysts/mL in 1 minute, 10 minutes, and 1 hour, respectively. For a sample of finished drinking water, the ELOD for *Cryptosporidium* oocysts was found to be 75 oocysts/mL, 7.5 oocysts/mL, and 1 oocyst/mL in 1 minute, 10 minutes, and 1 hour, respectively.
7. The experimental development plan included progressive instrumentation improvements that generated improved detection statistics but have not exhausted the scientific potential of the technique.
8. During this project, it was found that the existing sources for *Cryptosporidium* are limited, the sources for *Giardia* are unreliable, and that the sample quality and consistency, while perhaps sufficient for standard method testing, was not sufficient for techniques such as MALS.

In summary, this effort has demonstrated that MALS can detect and identify *Cryptosporidium* oocysts at perhaps useful levels in finished drinking water under controlled laboratory conditions. In addition, MALS can distinguish morphological changes associated with ozonation, heat treatment and excystation from untreated, viable oocysts.

RECOMMENDATIONS AND AREAS FOR FUTURE RESEARCH

1. Due to the limited scope of the project, it was not possible to perform as many repetitions of experiments as would have been preferred. This is particularly true in the area of MALS sensitivity and limits of detection. It is recommended that more experiments be performed to gain confidence in the performance of the system, especially at low spiking concentrations.
2. A limited set of matrix waters were tested in this project. In order to be able to apply this technology widely in the water industry, a much more extensive set of water matrices should be tested. If the first and most likely application of the MALS system is for post-treatment water, both in the plant and in the distribution system, the performance of the system should be measured against a wide and representative sample of types of post-treatment waters expected.
3. While *C. parvum* is an organism that is of high interest because of its resistance to standard treatment techniques, there are a number of other waterborne pathogens that are of concern in the treatment and distribution system, including for example *E. coli* and other enteric bacteria, aerobic spore formers such as *B. subtilis*, and *Giardia lamblia*. The MALS system should be challenged with these organisms to determine the ability for the system to speciate and distinguish between them and typical matrix background particles.
4. It would be of interest to the water industry to test the capabilities of the MALS system in a raw or source water environment. This project has demonstrated that while it is impractical to have a single instrument serve both treated and raw waters, a raw water system can be developed albeit, at a much lower volume scanned per time.
5. The experimental development plan included progressive instrumentation improvements that have not exhausted the scientific potential of the technique. New improvements could be applied that may provide as much as a 10 to 100 times improvement in performance in, for example, the Estimated Limit of Detection. Examples of these improvements and estimated potential improvement are: (a) running the water faster through the system, with a potential improvement of 2 to 5 times; (b) improved feature extraction techniques, with a potential improvement of 2 to 3 times; (c) improved electronics with a potential improvement of 2 to 4 times; and (d) improved optics with a potential of 2 to 3 times. Revisiting the types of experiments performed in this project would be useful once these new improvements have been achieved, in order to measure and quantify the improved performance of the MALS technology.
6. Upon further technological development, placement of MALS devices in strategic locations could allow a water agency, continuously and in real time, to: (a) detect incoming targeted organisms in source waters; (b) determine their removal through the plant; (c) determine the efficacy of the treatment process at the effluent of the plant; and (d) detect pathogens in distribution systems. In the current environment of high terrorist risk, and with the potential vulnerability of US water systems to terrorist attack, the applicability of the MALS technology should to be tested in a wider and more realistic environment such as a pilot project with several to tens of instruments placed in the treatment and distribution system.

CHAPTER 1
INTRODUCTION

THE NEED FOR REAL-TIME ON-LINE CONTINUOUS MONITORING

Rapid, reliable and accurate detection of microorganisms in drinking water is a longstanding and important need for the drinking water industry. *Cryptosporidium parvum* is of particular concern because it exhibits high resistance to disinfectants at the doses applied in water treatment plants, has low infectious dose, and no drug approved for prophylaxis or therapy is currently available. *Cryptosporidium* was responsible for one of the largest water supply contamination events in recent U.S. history in Milwaukee in 1993 (MacKenzie et al., 1994) when approximately 400,000 people became infected and about 100 people died from drinking contaminated water. With continued population growth and land development, and increased usage of recycled water, microbiological contamination incidences are likely to increase.

Current monitoring methods for microorganism contamination typically rely upon grab samples, transportation of samples to a laboratory, and labor-intensive laboratory techniques that often involve incubation such that results take hours or days. While these methods may suffice to meet existing compliance monitoring requirements, they provide no effective health benefit because by the time results are known, the water is already into the distribution system, delivered and likely consumed.

Current on-line monitoring methods such as turbidimeters and particle counters are commonly used as indicators to warn of potential contamination problems based on changes in turbidity or particle size distributions, respectively. However, because these techniques measure bulk properties (turbidimeters) or discriminate based on size only (particle counters) these techniques are inadequate for providing accurate or reliable monitoring for the presence of specific microorganisms such as *Cryptosporidium* (Nieminski et al., 2000). The Las Vegas *Cryptosporidium* incident in 1994 (Goldstein et al., 1996) is an example of an outbreak associated with a filtered water supply apparently meeting all appropriate standards for that time.

The purpose of this project was to provide a proof-of-concept evaluation and testing of multi-angle light scattering (MALS), a technology that has the potential to provide real-time, on-line, continuous monitoring for pathogens in drinking water. *Cryptosporidium parvum* was chosen as the test microorganism over other common waterborne pathogens for several reasons: it is not easily disinfected; it is not easily culturable; since 1993 it has been an organism of high interest to the water industry; it's size (4 μm to 6 μm) is in the middle of the size dynamic range for MALS at the wavelength used; and in its environmental (oocyst) form it is relatively inert and physically stable compared to vegetative bacteria such as *E. coli*.

The terrorist attack of 9/11/01 on the United States highlighted the increased security concerns for the nation's infrastructure, including water reservoirs, treatment systems, and distribution systems. Clearly, current standard techniques which rely upon grab samples are inadequate to monitor for intentional contamination events because they could happen without warning at unplanned times at unknown locations. The only viable means to provide timely alerts to an attack on the water infrastructure are real-time, on-line and continuous monitoring systems.

CURRENT METHODS FOR THE DETECTION AND MONITORING OF *CRYPTOSPORIDIUM*

Current techniques for *C. parvum* detection rely on various methods that have serious limitations. These techniques generally underestimate oocyst concentrations, possess low recovery efficiencies, and exhibit cross-reactivity to debris and other organisms. (c.f., Clancy and Hansen, 1999). As a consequence, a negative result does not guarantee that the water is free of *Cryptosporidium* contamination. In addition, the method requires highly technical and experienced personnel, only a limited number of samples can be processed simultaneously, it is a costly technique that requires excessive processing time, and analytical results are usually available after 24-72 hours from sample collection. Lengthy procedures for pathogen detection in water provide information "after-the-fact," thereby restricting water treatment and management options to prevent public exposure to contamination.

Some of the aforementioned limitations have been considered and overcome when implementing the United Kingdom regulations to monitor for *Cryptosporidium*. These systems are the first to operationally provide pseudo-continuous monitoring. The process includes 24-hour continuous filtration, filter shipping to a remote laboratory, filter elution, sample concentration, immunomagnetic separation (IMS), staining, and manual microscopic counting. Correspondence with several UK utilities indicates that the average cost per sample site is about \$120,000 to \$150,000/year, placing heavy financial burdens on the UK water companies to meet the regulation. In addition, results are not timely – typical turn-around time is three days. This delay greatly limits any public health benefit that the monitoring system might provide. In an unpublished case, Northwest Water Company (NWW) noted an unusually high level of *Cryptosporidium* oocysts in one of their test samples, corresponding to water that had passed their system three days earlier. Rather than issue a boil water order, since the contaminated water had already been delivered to customers, NWW issued an epidemiological alert to area medical authorities for increased cases of cryptosporidiosis within the next four to seven days. An elevated number of cases of cryptosporidiosis appeared as expected.

The currently recommended method for *C. parvum* detection in water – USEPA method 1623 (USEPA, 1996) – appears to be an improvement over its preceding Information Collection Rule (ICR; USEPA, 1996) method, although it still has several limitations. Method 1623 underestimates oocyst concentrations with recovery efficiencies ranging from 22 to 55% in spiked distilled water, and 14 to 70% in spiked raw water (USEPA, 2001), and exhibits cross-reactivity to debris and other organisms such as algae.

Another emerging method of detection of *C. parvum* in water consists of flow cytometry cell sorting (FCCS). While this method is more sensitive and rapid than the current methods discussed above, it still presents many limitations. As a strict laboratory method, detection is made of grab samples, which represent only a small fraction of the treated water. In addition, a 4-5 hour measurement time is needed in order to obtain useful results. This technique is complex and requires highly trained personnel. A few molecular-based, rapid screening methods have also been proposed, which involve sample concentration, and nucleic acid amplification with or without oligoprobe hybridization. Detection is achieved using different formats: liquid hybridization in a multi-well microplate (Baribeau et al., 1999); lateral flow chromatography (Kozwich et al., 1999); or melting curve analysis (Di Giovanni et al., 1999). Although more rapid and specific than the recommended methods such as Method 1623, these techniques still require collection of grab samples, making continuous and real-time monitoring unachievable. They also necessitate highly trained analysts and expensive equipment.

A REAL-TIME, ON-LINE CONTINUOUS METHOD: MULTI-ANGLE LIGHT SCATTERING (MALS)

Prior to the 1980's, light scattering had limited practical applications in biological particle identification because the optical inverse scattering problem had not been solved satisfactorily. Unlike the "forward scattering" problem, in which the scattered radiation from a particle is predictable, in principle, given sufficient information about the size, shape and refractive index of the particle, the inverse scattering problem is characterized by the determination of the physical properties of the particle from the scattered radiation.

A well-known solution to the inverse scattering problem, common in defense applications, is inverse synthetic aperture radar (ISAR). Radar is scattered from a moving target that changes its attitude relative to the radar. Scattered phase and amplitude information is collected, and an image of the target is reconstructed using signal processing techniques (Knott et al., 1985).

The analogous problem in optics is significantly more problematic because phase information is more difficult to obtain due to the short wavelengths involved (the visible light band is between 350 nm and 700 nm). Without phase information, a rigorous reconstruction of the scatterer, particularly a complex object, becomes untenable using standard techniques.

However, a solution to the inverse scattering problem using scattered amplitudes alone was achieved by Quist and Wyatt in the early 1980's (Quist and Wyatt, 1985). Using a technique called "strip maps," Quist and Wyatt demonstrated that it is possible to uniquely and rapidly identify a homogeneous and isotropic scattering particle using optical observables generated solely from the differential cross section (the angular dependence of the scattering amplitude) without explicit phase information. Because this technique relies upon simultaneous measurement of various scattered light angles, the technique is called multi-angle light scattering, or MALS. This technique has been demonstrated with various microparticles that produce coherent scattered light patterns with multiple nulls, including, for example, bacteria such as *E. coli* and *Staph. epidermidis* (Wyatt, 1968) and flyash (Sachweh et al., 1995).

In 1989, Wyatt and Jackson (Wyatt and Jackson, 1989) applied the MALS technique to microbiological particles in water. They were able to demonstrate that it is possible to discriminate between 12 distinct species of phytoplankton in seawater with a statistical confidence level of greater than 99%. This represented a significant result – inverse light scattering techniques used to rapidly identify very simple particles such as polystyrene latex (PSL) spheres could be extended to discriminate and identify complex biological particles with a high degree of confidence.

THE INSTRUMENTS

The multi-angle light scattering (MALS) instruments utilized in this project have the potential to resolve several of the limitations discussed above. MALS has a long and successful history of uses for other applications, including aerosol detection (Wyatt et al., 1988), biological assays for toxicants in water (Wyatt, 1989), macromolecular characterization (Wyatt Technology Corporation, Santa Barbara, CA), and most recently, the ability to detect, identify, and enumerate *Cryptosporidium* and *Giardia* in real-time (Quist, 2000). MALS uses laser light scattered from individual microparticles to identify the unknown scattering particles. A side stream of water passes through a "read-head," where particles pass through the laser beam and scatter light in all directions. The light is gathered at specific directions and subsequently used to uniquely identify the scattering particle (Quist and Ix, 2003).

MALS has many potential advantages over existing protozoan detection techniques. First, individual sensor read-heads are relatively small (including the electronics, about a cubic foot in volume), and can used at existing sampling points where a side stream of water (typically about 6 liters/hour) runs through the system. Second, measurements of the water take place directly – there are no processing steps (c.f. Method 1623) to decrease detection fidelity, propagate errors, or introduce sample contamination, uncertainty and error. Third, individual oocysts are detected essentially instantly, in less than 0.01 seconds as they pass through the laser beam in the read-head. Fourth, since the system can be automated, it can operate remotely with no operator or need for observation, and there is significant labor reduction and substantial potential cost savings. And finally, the technique can operate continuously, taking data as often as every five seconds, providing a detailed high fidelity time history of the water that has passed through the system. This detailed time history may be compared to other related continuous water quality data that may be available, such as particle counts and turbidity.

CHAPTER 2
MATERIALS AND METHODS

THE MALS MEASUREMENT SYSTEM

The multi-angle light scattering (MALS) system is an experimental measurement system that works in the following manner (c.f. Quist and Ix, 2003). A light source, typically a solid state laser, provides a coherent monochromatic illumination source. Typically red light (e.g. 685 nm wavelength) is used, although red is not required – any wavelength can in principle be used. A water stream flows across the laser beam such that individual particles are resolved as much as possible. As particles in the water pass through the laser beam, they scatter light in all directions. This scattered light is captured by optical detectors whose field of view is restricted to the region of the laser beam. Depending upon the size, refractive index, shape, internal structure and orientation, the intensity of the scattered light varies as a function of scattering angle (c.f. Bohren and Huffman, 1983). These optical scattering intensities are transformed into digital signals by optical sensors such as photomultiplier tubes or photodiodes and analog/digital electronics. The digitized signals are then compared with an existing library of light scattering signatures. The system software then makes a determination of identification – the particle is either positively identified as a target particle such as a *Cryptosporidium parvum* oocyst, or is rejected as "unknown." (Alternatively, each particle that passes through the system may be assigned a probability that that particle is *Cryptosporidium.*) This identification process, from particle passing through the beam to identification, takes less than 0.01 seconds.

Instrument Configuration

Four generations of real-time continuous instruments have been developed. The fourth generation ("Gen-4") is the most evolved and most successful instrument to date and was used for the experiments reported here. The Gen-4 MALS system is noted for having both the laser illuminator and the solid state optical detectors completely out of contact with the water stream. This significantly increases the required maintenance interval and reduces stability problems. Aspects of the Gen-4 system are discussed in the following and illustrated in Figure 2.1. A photograph of the Gen-4 read-head is shown in Figure 2.7. Numbers here correspond to Figure 2.1. The water stream is brought to the top of the instrument with a flexible tubing (1) which is attached to the top of a flow-through containment sphere (2). The fittings are all designed for smooth, laminar flow of water through all of these components. The Reynolds number for the water flow is well below the point of turbulent flow. The water exits from the bottom and flows through a needle valve (3). The needle valve is used to regulate the flow of water to a nominal 100 ml/min flow rate. This flow rate was chosen as a matter of convenience – flows of more than five times faster have been tested successfully with no degradation in performance of the system. The laser illuminator (4) creates a structured region of illumination by means of its focused beam, designed to limit the volume of the detection region.

As particles or microorganisms pass vertically through the containment sphere a portion of them pass through the laser beam and are illuminated. The laser light is linearly polarized and is at a wavelength of 685 nanometers. The scattered light, produced by particles passing through the beam, passes through the water and then through the containment sphere and into the surrounding

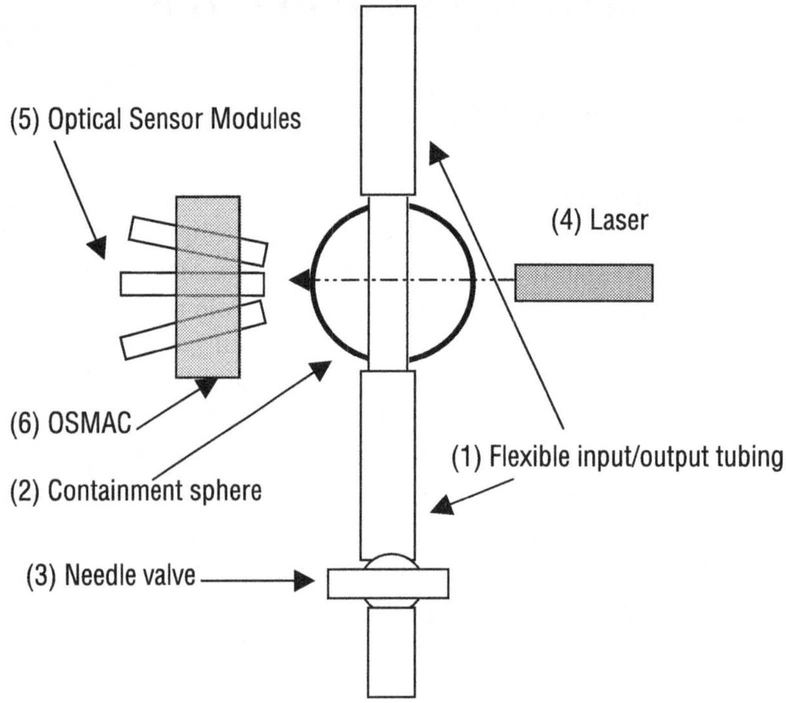

Figure 2.1 Schematic diagram of critical components of the Gen-4 MALS read-head

air. Optical detectors, called Optical Sensor Modules (OSMs) (5) are located in a rigid holder called the OSM Array Carrier, or OSMAC (6) . The positions of the OSMs in the OSMAC are set by observing the theoretically calculated positions for test spheres in the size range of 0.8 μm to 4 μm, placing the OSMs in regions that are likely locations for the detection of optical peaks and nulls in the optical scattering pattern (c.f. Wyatt, 1975). The positions used for this report were not tested to determine if they were optimal for *Cryptosporidium*. The OSMs have a narrow field of view that minimizes multiple particle interferences in the laser beam. There are 16 OSMs that form a region of intersection with the laser beam. This region is called the "Patrol Volume." The design Patrol Volume is about 1.5 mm × 1.5 mm × 0.1 mm, or 225 nanoliters. The OSMs form an image onto a solid state optical detector that is integrated with a low noise amplifier. Signals from the photodiode detectors are sent via a cable assembly to an analog-to-digital (A/D) board which digitizes the 16 signals at 12,500 Hz, thus taking data every 80 μseconds. These digitized signals are then sent to a central processing unit (CPU), which (a) determines if an "event" has taken place (a particle has passed through the laser beam), and (b) performs the identification of that particle. There are two choices for the identification: (a) the particle is matched with a target particle in the light scattering library and thus identified with that particle, or (b) it is called "unknown." An "unknown" particle is either a particle whose light scattering pattern is not sufficiently matched to a particle in the library, and thus not uniquely identifiable.

Individual instrument configurations (e.g. detector positions, electronic settings, laser beam geometry, laser power, etc.) directly affect the performance of the instrument because the

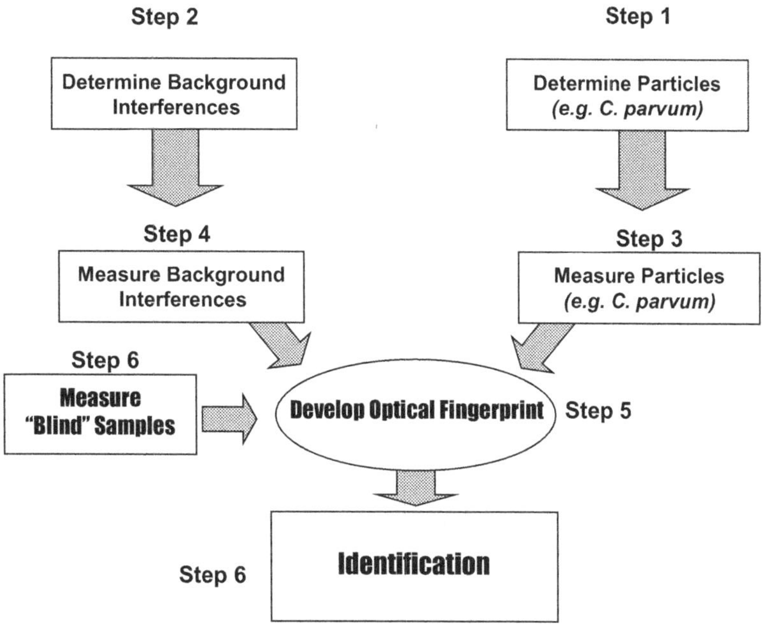

Figure 2.2 Schematic diagram of the development of optical fingerprints and subsequent identification process

Patrol Volume is different, the detector responses or positions are different, or the amplifier electronics are different.

Development of Optical Fingerprints

Since the application of the MALS technique to water quality monitoring is relatively new and probably unfamiliar, the generic data collection and measurement process will be described here.

The steps required to generate and test optical fingerprints (see Figure 2.2) in various backgrounds are as follows:

Step 1. Determine Target Particles

Prior to making "blind" measurements, the species of interest are determined. For example, in this project, the targets were: viable *C. parvum* oocysts, ozonated and inactivated forms of *C. parvum* oocysts, including empty oocyst shells, and polystyrene latex (PSL) spheres which are NIST qualified light scattering standards. Duke Scientific (Palo Alto, CA) PSL spheres were used throughout this project.

Step 2. Determine Background Interferences

Since measurements of these targets will generally take place in natural source waters or treated waters, there will be significant interference by background particles such as other non-target microorganisms and organic and inorganic particles. Interferences can either be specified,

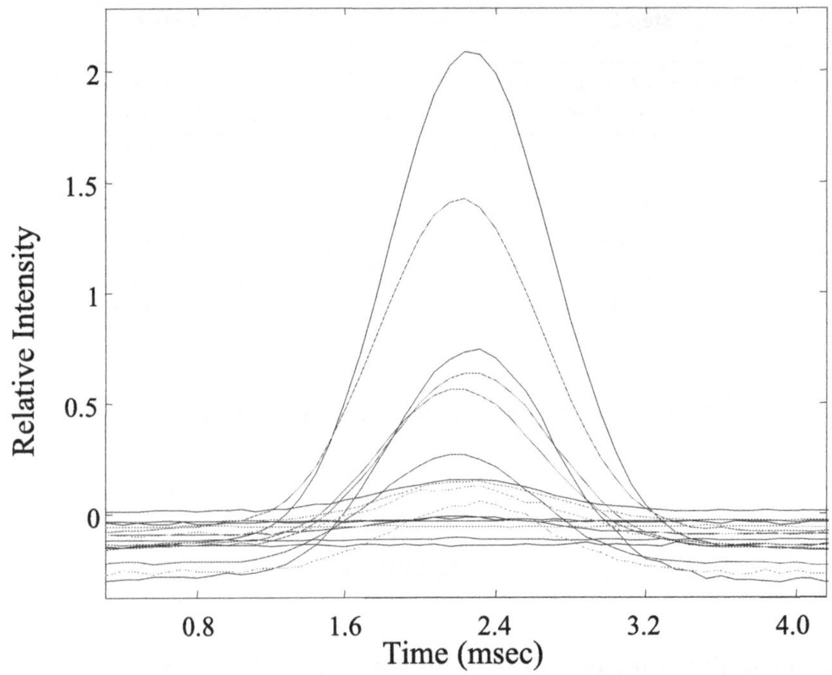

Figure 2.3 Sample of *Cryptosporidium parvum* oocyst event as seen in the raw data produced by the MALS Gen-4 instrument. Each curve corresponds to the response of a single detector channel.

as in distinct forms of algae or may be grouped together as "unknowns" in "representative" samples of background source and treated water.

Step 3. Measure Target Particles

Using low particulate water as described in this section as a background, a statistically significant number of measurements of individual targets are made. This number depends upon the desired fidelity of the measurements, but generally is in the several thousand to few tens of thousands range. Particle free water is necessary, because the error induced in this step is directly proportional to the fraction of particles that are erroneously considered to be target particles. For example, if 3% of the events are due to non-target (background) particles, the optical fingerprint produced by the software will have 3% of the points incorrect.

A measurement occurs when a particle passes through the laser beam and produces sufficient scattered light for the data collection software to determine that an "event has occurred." See Figure 2.3 for an example of a *C. parvum* oocyst event. For example, during this proof-of-concept research, several thousand measurements (events) of each species were made, and this number was sufficient for the identification process to be more than 90% accurate. The critical aspect is that enough measurements are made to: (1) ensure reproducibility of results and (2) ensure all reasonable physical variations and orientations of target particle are captured.

Step 4. Measure Background Interferences

The previous step is repeated for the background interferences, either as isolates such as other organisms or as a generic matrix water sample.

Step 5. Develop Observables

Once these data are collected, a statistical optimization process based on the multivariate analysis of variance (MANOVA) technique (c.f., Quist and Ix, 2003) is used to generate the optical fingerprints of the targets and background interferences. These observables are optimized for the specific target and interferences as determined in the previous steps.

Step 6. Validation and Identification

Once the optical fingerprints are generated (the "Test Mode"), blind experiments are performed to verify that the optical fingerprints are valid ("Validation Mode"). If the results in the Validation Mode are not satisfactory, the process can be repeated.

Instrument Evolution and Data Quality

There have been four generations of MALS instruments used as part of the development of a real-time, on-line continuous measurement system. These four generations are as follows.

Generation 1 System ("Gen-1")

The Gen-1 system was a batch system, the Wyatt Technology (Santa Barbara, CA) DAWN-B system, which used 20 mL borosilicate glass scintillation vials to contain small individual samples. See Figure 2.4 which shows a schematic top view of the instrument. Sixteen photodetectors were arrayed in single plane around the scattering center. The laser passes through the scintillation vial that contains the sample, and light is scattered into the detectors. The Gen-1 system was used prior to this project for the proof of concept of the MALS technology (c.f. Quist, 2000). The Gen-1 system demonstrated that it was possible to use multi-angle light scattering to differentiate pathogenic organisms of interest to the drinking water industry. The Gen-1 instrument does not provide real-time identification or continuous stream measurements.

Generation 2 System ("Gen-2")

The Gen-2 system was also a batch system like Gen-1. The major difference between in the Gen-2 system was that the detectors are arrayed in three dimensions and water flows continuously through this system with the water motion created by a stir bar. A photograph of the Gen-2 system under assembly is shown in Figure 2.5. The Gen-2 system was originally used to identify and distinguish 12 types of phytoplankton in filtered ocean water (Wyatt and Jackson, 1989). This system is an open structure flow-through system that used totally submerged components including the laser and fiber optic detectors attached to photomultiplier tubes. No real-time identification occurred with the Gen-2 system. The batch sample size for the Gen-2 system was 30 liters.

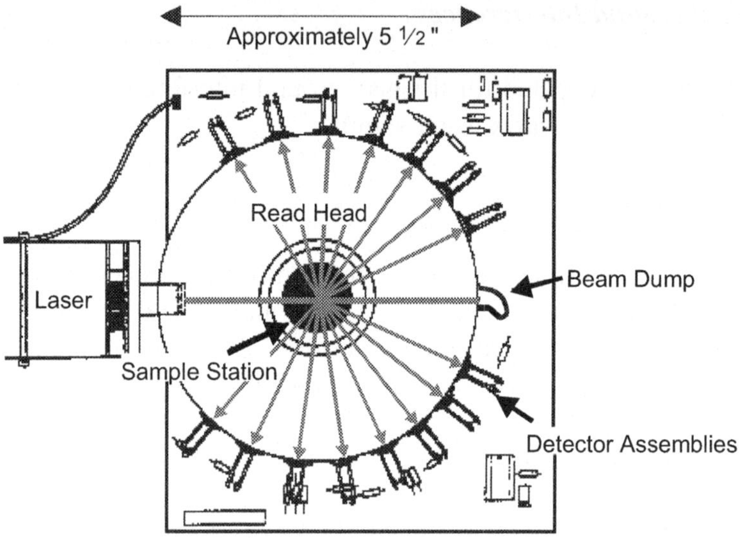

Figure 2.4 Schematic diagram of the Gen-1 MALS instrument, the DAWN-B system

Two aspects of the Gen-2 system limited its practical use. First, the sample size of 30 liters was very large. In order to develop high quality optical fingerprints, generally about 2,000 to 4,000 target events were required. High purity laboratory water was an aggressive solvent that leached particles out of its containment vessel and any other material in which it is in contact. These particles became contaminants to the optical fingerprints that were developed and decreased the fidelity of the fingerprints. Because of the high surface area of the containment vessel, it was observed that once the vessel became filled, background particles appeared after about 10 minutes of operation. Approximately 1.5 mL/min pass through the patrol volume. In order to gather about 2,000 events in 10 minutes, it required a very large spike of about:

$$\frac{2000 \text{ particles measured}}{(1.5 \text{ ml/min}) \times (10 \text{ min})} \times 1000 \text{ ml/liter} \times 30 \text{ liters} = 4 \times 10^{6} \text{ oocysts} \quad (2.1)$$

This very large number of oocysts and limited the number of experiments that could be performed during the project.

Second, the Gen-2 was not a practical system in terms of operations. It was a totally submerged system, and as such, had limited applicability as a field instrument, since it required the opening and closing of pipes for installation and maintenance.

Generation 3 System ("Gen-3")

The Gen-3 system was the first MALS flow-through system designed for side stream sampling. In this system several evolutionary changes were made. The fiber optic detectors were still submerged directly into water to minimize optical design, photodiode detectors replaced the photomultiplier tubes in the Gen-2 system, and the laser was taken out of the water.

Figure 2.5 Photograph of the Gen-2 system under assembly. A six-inch ruler is shown in the foreground for scale. The structure used to hold the fiber optic detectors (not shown) is on the left, and the enclosure for the laser is on the right. The complete assembly is fully immersed.

The Gen-3 system is shown in Figure 2.6. Water enters the read-head region on the top through Tygon tubing. The laser assembly, now out of the water, is on the right, and the fiber optic detectors, still immersed in the water are shown on the left side of the cylindrical scattering region. The black cylinder is six inches tall and three inches in diameter.

The Gen-3 system was the first continuous processing and real-time identification system. While the design of the Gen-3 system moved away from a submerged system towards a more practical system, the Gen-3 system was subject to several problems that limited the utility of the system. Optical sensor stability was affected by several problems: (a) Continued exposure to purified laboratory water caused slow degradation of the polarizers which formed the front surface of the detectors; (b) Degradation of the polarizers caused by disinfection techniques (a solution of 10% hydrogen peroxide was used as a disinfectant between runs of pathogens) also occurred; and (c) Macroscopic bubbles tended to form around the optical detectors, which had the potential for affecting the optics of the detectors. Water flow was non-laminar: the input and output sides of the read-head had tubes of inner diameter ⅜", while the scattering region was 3" in diameter. In addition, disinfecting the system required mechanical disassembly and reassembly of the major parts of the system and this resulted in additional data instabilities. The instabilities of the Gen-3 system made it very difficult to perform good experiments – after each time the system was disinfected, the system performance was changed in an unpredictable manner.

Figure 2.6 Photograph of the Gen-3 system installed in the laboratory

Generation 4 System ("Gen-4")

The Gen-4 system was developed to solve the deficiencies of the Gen-3 system. The detectors were taken out of contact with the water and water passed into, through, and out of the scatter region through a constant radius tubing.

A photograph of the Gen-4 system is shown in Figure 2.7. Water entered the read-head region through Tygon tubing and passed through the glass sphere. The detectors were out of the direct contact with the water. Water flowing through the system only contacted input and output ports and the glass sphere. The Gen-4 device represented the final step in moving from a batch MALS system to a dry flow-through system.

By changing the read-head from a large cylinder to a small glass cylinder, turn-around time for individual experiments went from about four to eight hours down to about one hour. This system showed excellent stability over periods of weeks, as shown in the Results and Discussion chapter section. Changes to the electronics and software also improved the quality of the data, as shown in the Results and Discussion section.

Operational Flow Diagram

A flow diagram illustrating the operations of the MALS Gen-4 system is shown in Figure 2.8.

The MALS system operated as follows (the numbers here correspond to the numbers in Figure 2.8).

1. Water passes through the read-head. The standard flow rate through the read-head was 100 mL/min. It should be noted that this was *not* the amount of water sampled per time by the read-head, since all of the water does not pass through the laser beam and all of the water is not visible to the OSM detectors.

Figure 2.7 Photograph of the Gen-4 system installed in the laboratory

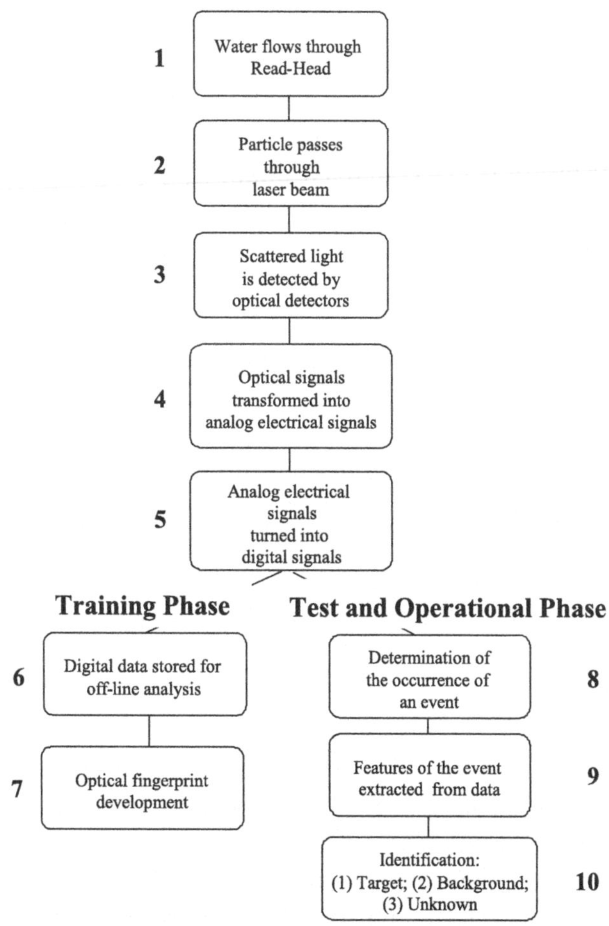

Figure 2.8 Operational flow diagram for the MALS Gen-4 system

2. A particle in the water passes through the laser beam in a region that is in view of all 16 optical detectors. The laser beam cross section is typically 1.5 mm (wide) × 0.1 mm (high).
3. Light scattered by the particle is gathered by all of the detectors.
4. The photodetectors transform and amplify the analog optical signal into an analog electrical signal.
5. These analog signals are digitized by a standard commercial A/D board at a rate of 12,500 Hz for all 16 channels of data.

Then, depending upon whether the data were being collected during the Training Phase, when the optical fingerprint was developed, or the Test and Operational Phase, when water is being testing, the data is treated as follows.

Training Phase

6. The data were stored for off-line analysis.
7. The optical fingerprints were developed as described earlier.

Test and Operational Phase

8. The data were tested for the occurrence of an Event.
9. If an Event has occurred, features of the Event are selected and compared to the optical fingerprints developed in the Training Phase.
10. An Identification is made: The Event, generated by a particle passing through the laser beam, is classified as a (a) Target; (b) Background particle, that is a particle expected in the matrix water; or (c) Declared as "unknown."

Comparison Between Traditional Methods (e.g., Method 1623) for *Cryptosporidium* Monitoring and MALS

Because MALS is a new method which continuously measures a sample rather than collects and analyzes a sample, it is useful to compare the characteristics of the standard methods, such as Method 1623, with MALS. Table 2.1 shows a comparison between the MALS technique and standard grab sampling methods for several relevant parameters.

Performance Parameters

Performance parameters for the MALS method are defined here.

Effective Patrol Volume Rate

The Effective Patrol Volume Rate (EPVR) is the amount of water fully patrolled (sampled) per unit time by the MALS read-head. The Effective Patrol Volume Rate is determined by: (a) the speed of the water flowing through the read-head; (b) the laser beam geometry (cross section); (c) the field of view of the 16 optical detector set; and (d) the size and shape of the target particle of interest. Typically, 100 ml/min flowed though the Gen-4 read-head, which in the ⅜" cross section generated a water speed of about 10 cm/sec. This flow speed was kept constant during this effort.

Table 2.1
Comparison of some method characteristics between standard grab sample methods, such as Method 1623 and MALS

Characteristic	Standard grab sample	Multi-angle light scattering
Sample Collection	Single point in time, or short period of time.	Continuous.
Sample Transportation	Requires physical transportation to lab.	Analysis takes place on-site at point of sample collection
Sample Volume	One-time sample, typically 10 liters to 1000 liters.	Sample is monitored continuously – currently 100 ml/min flows through the system, but only about 10 to 120 ml/hour is fully monitored by the system, depending upon particle size and shape.
Recovery	The ratio of the number of oocysts captured and counted at the end of the process to the number that have been spiked into the sample.	No physical capture takes place – oocysts flow through the system and are identified as *Cryptosporidium* or are missed. An analogous value to recovery would be the ratio of corrected identified oocysts to the number that have passed through the system's identification process.

For a fixed instrument configuration (e.g., laser geometry, water flow speed and detector configuration), the size and shape of the particles affected the Effective Patrol Volume Rate as follows.

In order for the system to recognize that a particle has passed through the Patrol Volume, most of the detectors have to register a signal that is above the noise floor (c.f., Figure 2.3). Larger particles will generally scatter more light than smaller particles. Since the profile of the laser beam is essentially gaussian, and the noise floor is fixed and independent of particle size, a sample with a given concentration of larger particles will generate more events per time than a sample with the same concentration of smaller particles. The shape of the particles has an effect, too. Asymmetric particles will generate different light scattering patterns depending upon orientation in the laser beam. If certain orientations direct scatter in regions that are not covered by the detectors, the particle will not be detected by the read-head. Therefore, larger particles will have higher Effective Patrol Volume Rates than smaller particles and symmetric particles will have higher Effective Patrol Volume Rates than an asymmetric particles of the same average diameter.

Identification Rate

The Identification Rate (IDR) is defined as the ratio of correctly identified *Cryptosporidium* oocysts to the number of oocysts detected as passing through the system. The IDR is measured in purified laboratory water using a spike of known concentration. A perfect IDR would be 100%.

False Negative Rate

The False Negative Rate (FNR) is the ratio of the number of *Cryptosporidium* oocysts that are missed by the system – incorrectly called "unknown" – to the total number of oocysts detected as passing through the system. The FNR and the Identification Rate (IDR) are related by: IDR = 1 – FNR. A perfect FNR would be 0%.

False Positive Rate

The False Positive Rate (FPR) is defined as the ratio of matrix particles incorrectly identified as *Cryptosporidium* to the number of matrix particles detected. The FPR is measured using a blank (unspiked) matrix sample with the assumption that there are no naturally occurring *Cryptosporidium* oocysts in an unspiked matrix sample. A perfect FPR would be 0%.

Interference Factor

The Interference Factor (IF) is a measure of the decrease of the Identification Rate (or increase of False Negative Rate) in a real water matrix compared to an ideal purified laboratory matrix, i.e., it is a measure of the interference caused by the real water matrix. Since the MALS technique relies upon a single particle being present in the Patrol Volume at a time, it is expected that the Interference Factor will be at least dependent upon matrix particle content, as well as other possible factors such as oocyst agglomeration or chemical effects. The Interference Factor is typically measured as a percentage between 0% (no interference from the background matrix), and 100% (the matrix completely interferes with the measurement and no identification can occur).

Estimated Limit of Detection

The Estimated Limit of Detection (ELOD) is the lowest concentration at which the signal from the MALS system is at least three standard deviations above the background noise. Because most of the experiments to determine Limit of Detection were performed with large jumps in oocyst concentration, the Limit of Detection (LOD) was estimated, as opposed to directly measured, by either extrapolations or interpolations of measured results.

There are effectively two major ways to alter the performance parameters listed here. First, the instrument configuration can be changed, via optical or mechanical means. Second, the analysis software can be changed. For example, the software can be modified to allow more particles to be accepted as events by changing the number of detectors that are required to define that an event has occurred. A significant portion of this project was dedicated to finding the best physical and analytical settings in order to optimize these performance parameters.

Calibration and Stability

The optical fingerprints which form the basis of the analysis using MALS are based on comparisons of the optical scattering properties of a target versus other targets, or a target versus particles in a matrix background. Since these are relative measurements, no absolute calibration of the system is necessary. However, the system must remain stable over the periods of the experiments in order to compare results from day to day.

Table 2.2
Illustration of bacterial and particle contamination of a *Cryptosporidium* oocyst sample

Suspension	Lot no.	Concentration oocysts (#/mL)	Viability (%)	Concentration bacteria (#/mL)	Concentration particles (#/mL)
C. parvum	10122-27	2.7×10^6	82.0 ± 6.1	1.7×10^7	1.1×10^7

System stability was measured as follows. An optical fingerprint consisting of 1 μm diameter PSL spheres, 2 μm diameter PSL spheres, and 4 μm diameter PSL spheres (at a concentration of either 250/mL or 500/mL) in a background matrix of purified laboratory water (LW) was developed. Every few days, depending upon the number of experiments run per day, the optical fingerprint was tested using 4 μm diameter PSL spheres. If the Identification Rate for the 4 μm spheres was greater than 90%, then the system was considered to be stable. Results of such stability testing are shown in the Results and Discussion section.

SOURCES AND HANDLING OF *C. PARVUM* OOCYSTS

The *C. parvum* oocysts used during this project were produced by the Sterling Parasitology Laboratory at the University of Arizona, Tucson. The oocysts were from the Iowa isolate originally obtained from Harley Moon (National Animal Disease Center, Ames, Iowa). This isolate was selected because it is well characterized and has been used in numerous studies, including analysis of new methods of isolating *Cryptosporidium* oocysts and *Giardia* cysts (Clancy et al., 2000).

Oocysts were produced by infecting Holstein calves. The fecal output was collected, screened through sieves and concentrated by centrifugation. Oocysts were then isolated from feces by discontinuous sucrose gradients followed by microcentrifuge-scale cesium chloride gradients. Purified oocysts were stored at 4°C in 0.01% Tween 20 solution containing 100 units of penicillin, 100 μg of streptomycin, and 100 μg of gentamicin per ml to retard bacterial growth (Clancy et al., 2000). To avoid issues regarding the possible physical or morphological change in oocysts due to aging, only oocysts less than one month in age (from shed date) were used in testing.

Samples thus prepared were measured in the Gen-2 and Gen-3 MALS devices. It was observed that the variance in the light scattering results was larger than expected in these devices, so a more careful look at the sample was undertaken. In addition to *Cryptosporidium* oocysts, the samples contained high numbers of bacteria (dead or alive) and other detritus that is a residual of the purification process. Enumeration by standard microscopy that was performed by Clancy Environmental Consultants (St. Albans, VT) of a *Cryptosporidium* sample received by Point-Source in March 2001, for example, yielded the approximate particle count as shown in Table 2.2. For this particular lot, the ratio of "other" particles (bacteria and detritus) to oocysts was more than 10 to 1.

Since the quality of a particular optical fingerprint is a direct function of the purity of the sample of the target microorganism, two separate techniques were undertaken to reduce the particulate contamination of the *Cryptosporidium* oocyst samples. First, the samples were subjected to a second, more rigorous cesium chloride gradient. Second, the samples were sent to the Wisconsin State Laboratory of Hygiene to be separated using flow cytometry. Several of each of these samples were provided and tested using MALS.

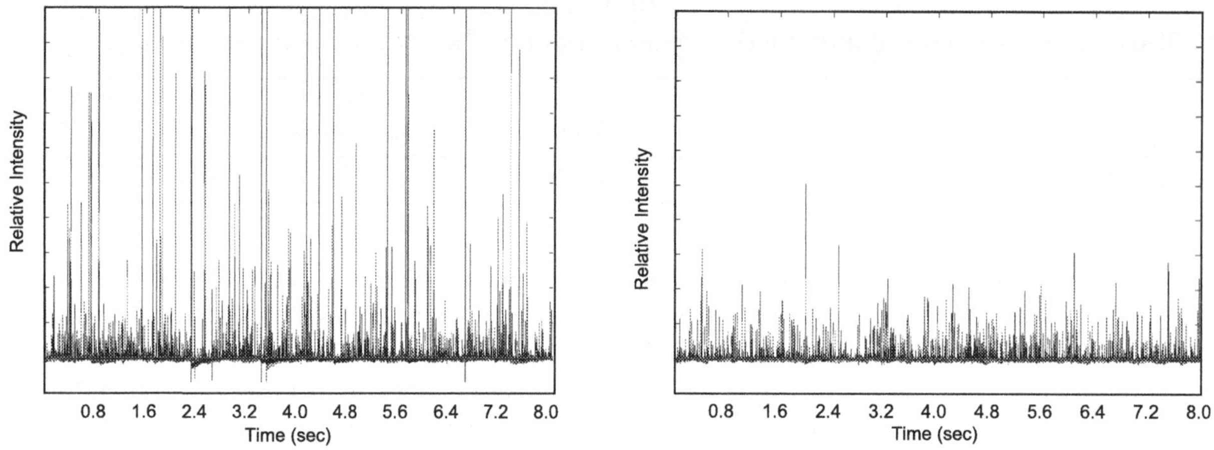

Figure 2.9 Comparison of typical MALS light scattering patterns from two *Cryptosporidium* oocyst samples, with the final purification step using cesium chloride (CsCl) gradient (left) and flow cytometry cell sorting (right)

Figure 2.9 shows the difference in the MALS light scattering patterns between a cesium chloride (CsCl) gradient sample and a flow cytometry sample. The CsCl samples contained a residual white precipitate which became suspended upon sample preparation, which included vortexing for two minutes. This precipitate was clearly an interfering source as seen in the left hand figure in Figure 2.9 and generally rendered the CsCl samples unacceptable for MALS optical fingerprinting. Several events in this 8 second sample exceeded saturation and were indicative of large (greater than 10 μm diameter) particle contamination. In contrast, the flow cytometry cell sorting (FCCS) samples were visibly clean as seen in the right hand figure in Figure 2.9. This difference is reflected also in Figure 2.10.

In addition to raw MALS data shown in Figure 2.9, the two oocyst samples were also processed analytically versus the same Weymouth Plant Effluent (WPE) drinking water sample to observe the ability of these samples to form good MALS optical fingerprints. The results of this analysis is shown in Figure 2.10. It is obvious by inspection of the results in Figure 2.10 that the separation for the FCCS sample was superior to the CsCl sample, and this was confirmed by the Wilk's lambda from the multiple analysis of variance (MANOVA) analysis (c.f. Cooley and Lohnes, 1971) for each case. The Wilk's lambda for the CsCl-WPE case was 0.8402 and for the FCCS-WPE case was 0.3980. A smaller Wilk's lambda indicates better MANOVA discrimination.

Based on these results, all measurements for this project used the FCCS samples provided by the University of Arizona and processed by the Wisconsin State Laboratory of Hygiene.

CONTINUOUS CIRCULATION SYSTEM

In order to develop a high quality optical scattering pattern of the *Cryptosporidium* oocysts, a nearly particle-free background is required. The PointSource laboratory water system (see description in the "Source and Generation of Purified Water" section), which consisted of a reverse osmosis unit continuously circulating through a series of filters, generated nearly particle-free water. Based on MALS light scattering data, typical particle concentrations were estimated to

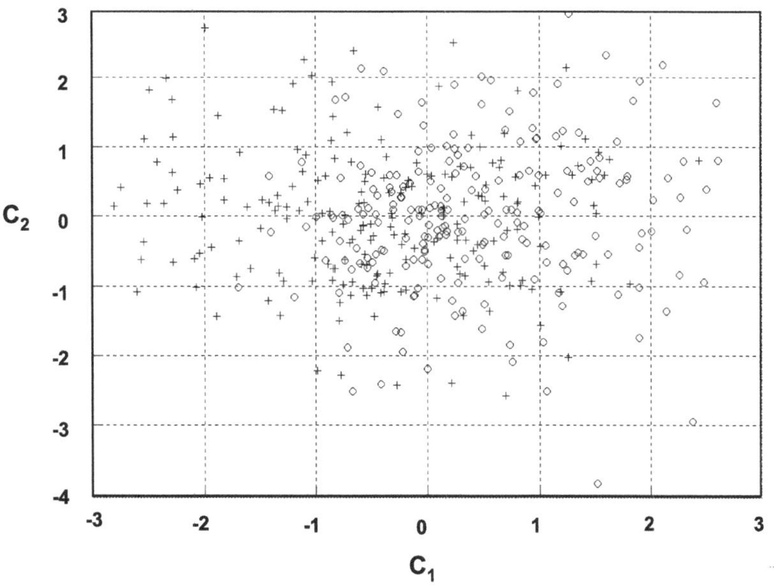

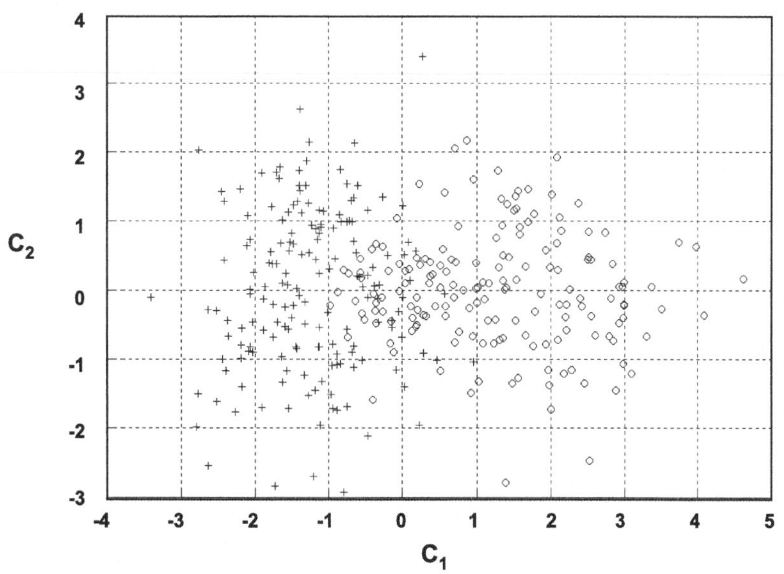

Figure 2.10 Output of MANOVA analysis routine, illustrating the difference in MALS optical fingerprinting between the CsCl purified sample versus WPE (top) and the FCCS purified sample versus WPE (bottom). Crosses are *Cryptosporidium* and circles are background particles in WPE.

be less than 10 particles per mL of size 500 nm and larger. This estimate was verified by particle counts for laboratory water (using a Met-One Water Grab Sampler Model WGS-267, Pacific Scientific, Grants Pass, OR) of less than 10 particles of size 2 μm diameter and larger per mL.

It was initially planned to use a continuously pumped, closed circulation system to make the measurements of spiked and unspiked water samples. The first tests of such a circulation system failed due to excessive particle contamination. Peristaltic pumps and diaphragm pumps were tested using high quality low particulate laboratory tubing, but neither system was satisfactory. The continuous circulation system was abandoned and instead, flow through was achieved using an elevated 8.5 liter reservoir, with a gravity fed tube through the MALS read-head (see Figure 2.11). This mimics common configurations in water treatment plants where taps are provided, for example, for continuous sampling of turbidity and particle counts. Water speed, and therefore sampling rate, was controlled by means of a manual valve located beneath the read-head. The standard flow rate through the system was 100 mL/min.

SPIKING METHODS

A large water purification system (see "Source and Generation of Purified Water" section) provided low-particulate purified water for cleaning and for some analyses. This water was introduced into the jar by means of a probe protruding through the top. Drinking water and other matrices were added to the cleaned jar by pouring from flasks. The jar was cleaned by flushing water in through the probe and out through the lower valves prior to analysis. The system was checked with a data collection run (the blank run). After passing the blank criteria, a known amount of the desired matrix was added to the jar (usually 3 liters). This was tested either for background characterization, or particle species (*Cryptosporidium*, etc.) that were spiked into the matrix using Eppendorf air-displacement mechanical pipettes.

Target particle species in this project included various sizes of Duke polystyrene latex (PSL) spheres, *Cryptosporidium parvum.* The microbiological species were stored at 4-8°C for up to four weeks after receipt. After four weeks, the samples were no longer used. Flow cytometry purification includes cell counting, so the concentrations of these preparations were generally well known.

Typical experiments involved testing the particle species of interest in both a very low particle background (purified water or, potentially, various ultrafiltered liquids) and drinking water. Unspiked drinking water was also tested for background characterization.

Cryptosporidium samples were vortexed for at least 2 minutes prior to being measured into the jar. The spiked jar was mixed with a stir bar and stir plate for 6 minutes prior to opening the lower valve to 100 ml/min and acquiring data.

The accuracy of 4 μm Duke sphere spikes can be assessed using a Pacific Scientific Met-One particle counter. There was no way of assessing the accuracy of other spikes, or the accuracy of the stated concentrations of most of the starting biological materials. The variability levels of the spikes were best assessed using the historical 4 μm particle counter results, but this did not include variability or errors in the concentration determinations made out-of-house by flow cytometry or plate counting.

SOURCE AND GENERATION OF PURIFIED WATER

Very low-particulate water was the starting point for the optical fingerprinting process, as well as for the oocyst physiological difference testing. Very low-particulate water was produced as

Figure 2.11 Photograph showing the basic experimental set-up. A water sample (spiked or unspiked) was introduced into the MALS read-head via gravity from the 8.5 liter reservoir. Effluent from the read-head (pictured is the Gen-3 read-head) is collected in a collection reservoir below the read-head for disinfection and disposal.

follows. Tap water was pre-conditioned with an industrial water softener (US Filter, Palm Desert, CA) to remove dissolved solids followed by a charcoal filter (US Filter, Palm Desert, CA) to remove residual chlorine. Large particulates were then removed with a 5 μm filter. The conditioned water was then treated by a RO system (US Filter, Palm Desert, CA) and stored in a large 125 gallon reservoir. The post-RO water was continuously circulated with a system that provides UV disinfection on the input and output side of the reservoir and three polishing filters of consecutive smaller bore sizes: 0.2 μm, 0.1 μm, and 0.05 μm. Water produced in this manner was typically found to have particle densities of less than approximately 10 particles/cm^3 for particles greater than 500 nm in diameter.

SOURCES OF MATRIX WATERS

Four water samples were collected from a total of five sources. The water samples were collected in pre-washed 10 liter Nalgene carboys. Samples were collected at the following sources:

1. Effluent from the Metropolitan Water District Weymouth Filtration Plant (La Verne, CA).
2. Colorado River water collected by the Metropolitan Water District from the Diemer Filtration Plant.
3. State Water Project water collected by the Metropolitan Water District from Live Oak Reservoir in La Verne, CA.
4. Ground water from the Orange County Water District, collected at Prado wetlands, well ID "CBWM 0600664."
5. Ground water from the Irvine Ranch Water District, collected at the corner of Redhill Ave. & Barranca Parkway, Santa Ana.

The Metropolitan Water District Weymouth Filtration Plant in La Verne, CA uses a blend of raw source waters from the Colorado River Aqueduct and the California State Water Project.

Water samples from post-flocculation/sedimentation and pre-filtration were originally considered as potential samples. However, it was determined that water collected post-flocculation/ sedimentation and pre-filtration from the Weymouth plant would likely have particle concentrations exceeding 10,000/mL, therefore, this water was not collected and tested. The MALS system, in its current configuration, was not expected to operate with particle densities exceeding this value, since the sample region will nearly always contain more than one particle at a time, confusing the light scattering signal from an individual particle. However, geometry configuration modifications and a narrower laser beam could improve the performance of MALS at the expense of sampling less water per unit time.

LIGHT SCATTERING STANDARDS

The light scattering standards used in the project were polystyrene latex (PSL) spheres from Duke Scientific (Palo Alto, CA). These spheres are NIST-traceable particle size standards. The PSL spheres were highly uniform and have a refractive index in the red band of 1.59, which was higher than the average refractive index of *C. parvum* oocysts. It is this difference in refractive index that causes the well-known effect with particle counters to mischaracterize *C. parvum* oocysts as particles that are roughly 2 µm in diameter. Spheres of diameter 1 µm, 2 µm and 4 µm were used during this project, bracketing in size and integrated scattered power the *C. parvum* oocysts. These spheres were used to test the reproducibility of the MALS measurements and provide consistent light scattering response for measurement of instrument stability.

PARTICLE COUNTING

As an independent means of monitoring water quality, a MET-One Water Grab Sampler (Model WGS-267) particle counter was used to determine the particle count in unspiked matrix water samples and matrix water samples spiked with 4 µm PSL spheres. The lower size limit for the MET-One was specified to be 2 µm, but spiking tests indicated that less than 50% of the 2 µm particles were typically counted by the system used. The MET-One was not used for *Cryptosporidium* spikes, as the manufacturer advised against using dilute (3.5% to 10%) hydrogen peroxide – the disinfectant used during this project – to clean the system. The system was factory calibrated, and was checked with fixed spiking experiments and found to be 90% accurate up to about 6000 particles (4 μm PSL spheres) per mL.

WATER CHEMISTRY METHODS

Polypropylene carboys (10 L) were washed in liquinox detergent, rinsed with tap water and deionized water, air dried, and sterilized by autoclaving at 121°C for 15 min. Filter effluent from the F. E. Weymouth filtration plant was collected from a dedicated sample port; carboys were rinsed with 250 mL of filter effluent water prior to sample collection. Water samples were also collected in 2 L and 100 mL bottles for chemical analyses. Samples were stored at 4°C and were transported on ice.

Water temperature was measured using a calibrated mercury filled thermometer, adjusted to the zero mark and then completely immersed in the water sample for 2 min. The temperature was recorded immediately after sample collection. The pH was measured using an Orion EA 920 ion analyzer that was calibrated daily with standard pH 7.00 and pH 10.01 buffers (Beckman Coulter, Inc., Fullerton, CA) following the manufacturer's instructions. Water samples were equilibrated to approximately 25°C before measuring the pH. The turbidity of the water samples was measured within 8 h of sample collection. If not measured immediately, samples were stored in the dark at 4°C until analysis. Turbidity was measured using a Hach 2100 AN Turbidimeter (Loveland, CO), which was calibrated with formazin turbidity standards following the manufacturer's recommendations. A cuvette was filled with sample water, and sonicated for 5 sec. The cuvette was wiped with a lint-free cloth to remove dust and fingerprints prior to analysis. The peak measurement was recorded as nephelometric turbidity units (NTU).

Free chlorine residual was measured using N,N-diethyl-p-phenylenediamine (DPD) powder (PermaChem Reagents) and a Hach DR/700 colorimeter (Loveland, CO) equipped with a 52.01 module according to the manufacturer's instructions. Chlorine in the sample immediately reacts with the DPD indicator to form a red color, the intensity of which is proportional to the chlorine concentration. Free chlorine was analyzed immediately upon sample collection as follows. A 10 mL cell was filled with sample water and used as a blank to zero the instrument to 0.0 mg/L. A second 10 mL cell was filled with sample water and the contents of one DPD reagent package added. The cell was capped, shaken for 20 sec, and placed in the instrument's cell holder. The free chlorine concentration was recorded as mg/L within 1 min of reagent addition.

Total organic carbon (TOC) was measured using the UV persulfate oxidation method described in *Standard Methods for the Examination of Water and Wastewater* (APHA, AWWA, and WEF, 1998) section 5310 C.

Total dissolved solids (TDS)/total filterable residue were measured using the method described in *Standard Methods for the Examination of Water and Wastewater* (20th edition) section 2540 C.

Total suspended solids (TSS) were measured using the method described in *Standard Methods for the Examination of Water and Wastewater* (20th edition) section 2540 D.

PHYSIOLOGICALLY ALTERED OOCYSTS

Flow cytometry sorted and enumerated oocysts of the Iowa isolate (lot number 021105-13) that were received from the Wisconsin State Laboratory of Hygiene and disinfected by exposure to UV light, ozone, and heat. In addition, oocysts were excysted to release their sporozoites. The infectivity of these treated oocysts was measured using in-vitro cell culture in HCT-8 cells combined with a reverse transcriptase polymerase chain reaction (RT-PCR) assay as described previously (Rochelle et al., 1997; Rochelle et al., 2002). The ID_{50} of untreated control oocysts was 128 which is within the usual range of infectivity for this isolate (95% confidence interval of the ID_{50} was 52 – 154 based on 13 replicate analyses; Rochelle et al., 2002). All treated oocysts were stained with a fluorescein-labeled antibody (Crypt-a-Glo, Waterborne, Inc., New Orleans, LA) and examined by epifluorescence microscopy. Untreated control oocysts displayed the typical microscopic appearance and staining characteristics of intact, round oocysts, 5 µm in diameter with smooth, intensely stained edges and lighter stained centers.

Oocysts were incubated at 80°C for 45 min to render them non-infectious. This resulted in ≥3.72 $\log_{10}$ inactivation based on the cell culture assay. Microscopic examination revealed typically stained oocysts with slightly irregular edges.

Oocysts were exposed to UV light (44.6 mJ/cm^2) using a medium pressure UV lamp mounted above a 6.4 cm inside diameter collimating tube (Calgon Carbon Corporation, Pittsburgh, PA). The sample was continuously mixed in a 60-mm polymethylpentene petri dish using a stir bar and magnetic stirrer during UV exposure. Incident UV irradiance was measured using an IL1700 radiometer (International Light, Newburyport, MA) with mathematical correction and was confirmed utilizing chemical actinometry to calibrate the dosages delivered to the oocyst suspension. More detailed descriptions of the UV apparatus and the methods used to measure UV dosages have been published previously (Linden and Mofidi, 1999; Mofidi et al., 2001). Exposure to 44.6 mJ/cm^2 of medium-pressure UV light resulted in 3.5 $\log_{10}$ inactivation and a microscopic appearance identical to control oocysts.

Oocysts were also exposed to ozone in a simple small volume batch reactor at room temperature. Ozone was delivered into a solution of 0.05 M phosphate buffer (pH 7.4) by an Osmonics V5-0 ozone generator (Minnetonka, MN) with 100% oxygen gas as the source, to achieve a dissolved ozone concentration of approximately 10 mg/L. An aliquot of this stock ozone solution was then added to 33 mL of an oocyst suspension and incubated at room temperature for 60 min to obtain a CT of 160 mg min/L. Ozone was measured using an indigo colorimetric method (Method 4500-O_3; APHA, AWWA, and WEF, 1998) and residual ozone was quenched by the addition of sodium thiosulfate. Exposure to ozone resulted in ≥3.72 $\log_{10}$ inactivation based on the cell culture assay and the oocysts had irregular edges but typical fluorescent staining characteristics when examined by fluorescence microscopy.

For in-vitro excystation, oocysts were pelleted by centrifugation at 4,000 rpm for 15 min at 19°C and resuspended in 1 mL of sterile acidified phosphate buffered saline (pH 3.8) and incubated at 37°C for 30 min. The oocysts were pelleted again by centrifugation and resuspended in 100 µL of Hanks balanced salt solution (HBSS; Sigma Chemical Co., St. Louis, MO). Following addition of 200 µL of 1% bile salts in HBSS and 50 µL of 0.04% sodium bicarbonate, the oocysts were incubated for 17 h at 37°C. Excysted oocysts were observed by light microscopy. The excystation efficiency was 83.3% and the level of infectivity of these oocysts was similar to the control oocysts. Microscopic examination of these oocysts following fluorescent staining revealed mostly broken oocysts with irregular edges and intense staining.

OOCYST ENUMERATION

Oocyst stock preparations were enumerated using a Bright-Line hemacytometer (Hausser Scientific, Horsham, PA). All tubes and pipette tips were rinsed with 0.01% Tween-20 to avoid adherence of the oocysts to the plastic to maximize the accuracy of enumeration. With the coverslip in place, 10-12 μL of diluted sample was transferred to each chamber of the hemacytometer and oocysts were enumerated following the manufacturer's instructions using a Nikon Optiphot-2 microscope under 200 × magnification. Dilutions were prepared to ensure that the total number of oocysts in the squares that were counted was in the range 200 – 500. All enumerations were performed in triplicate. The final oocyst concentration was calculated as the average number of oocysts in five squares multiplied by 10^4 and the dilution factor.

Oocysts were also enumerated by immunofluorescence microscopy following staining with a fluorescein-labeled antibody (Crypt-a-Glo; Waterborne, Inc., New Orleans, LA). The initial oocyst stock was diluted in PBS to yield approximately 20 – 200 oocysts per well. A 100 μL aliquot of diluted oocyst suspension was spread over each well of a Superstick slide (Waterborne, Inc., New Orleans, LA) and allowed to dry overnight in a light-protected plastic box containing desiccant. The FITC-conjugated antibody was diluted according to the manufacturer's recommendations using the supplied diluent and 75 μL was spread over the entire well. Slides were incubated at 37°C in a dark, humid chamber for 30 min. The antibody solution was removed by wicking-off with a paper towel and the wells were washed with 75 μL of PBS, followed by sterile deionized water and excess liquid was again removed by wicking. Slides were dried briefly in a desiccating box and then mounted in 1 drop of mounting medium (Meridian Diagnostics, Cincinnati, OH) under a 22 × 50 mm coverslip. The oocysts in each well were enumerated under 200 × magnification using a Nikon Optiphot-2 fluorescence microscope with excitation and emission wavelength filters of 450 – 490 nm and 515 nm, respectively. Oocyst concentrations were reported as the mean ± standard deviation of 5 – 12 replicate well counts.

OOCYST RECOVERY BY IMMUNOMAGNETIC SEPARATION

In some instances it was necessary to concentrate oocysts from larger volumes of water by immunomagnetic separation (IMS) prior to enumeration. In these cases, a modified version of the IMS protocol described in USEPA Method 1623 was used. The initial volume of oocyst-spiked water sample was determined using a 250 mL centrifuge tube. Following centrifugation at 4,300 rpm for 16 min at 19°C in a Sorval RC5 centrifuge, the supernatant was reduced to 30 mL using a vacuum aspirator and then reduced further to 5 mL using a Tween-20 (0.01%) rinsed glass pipette. The tube was vortexed for 1 min and the contents transferred to a Leighton tube containing 1 mL each of Buffers A and B (Dynal, Lake Success, NY). The 250 mL centrifuge was rinsed twice with 4 mL of deionized water and the rinsate added to the Leighton tube to yield a final volume of 10 mL. The sample was incubated at room temperature on a rotary mixer for 60 min and the oocysts were recovered by IMS as described in Method 1623. Following the acid disassociation step, oocysts were transferred to a 1.5 mL tube containing 5 μL of 1 M NaOH to neutralize the acid. The volume was increased to 1 mL with sterile DI water. This suspension was then enumerated by immunofluorescence microscopy as described above.

CHAPTER 3
RESULTS AND DISCUSSION

MALS GENERATION 4 INSTRUMENT RESULTS

The MALS Gen-4 instrument, described in detail in Chapter 2, was used for the results described in this section. This system provided the most consistent and reliable results due to the engineering changes: the detectors were out of direct contact with the water and several electronic and optical improvements were made over previous generations of instrumentation.

Results are shown in the following order, representing increasingly difficult tests for the MALS system:

- Polystyrene latex (PSL) spheres spiked into purified lab water (LW). Standard light scattering sphere measurements were made frequently and were used to test the performance of the system after hardware or software modifications and on an on-going basis for stability. The purified laboratory water results generally represented ideal measurement conditions and best possible MALS performance for the given configuration of the instrument.
- *Cryptosporidium* oocysts spiked into purified lab water (LW). These measurements represented the highest performance of the MALS system – for its particular configuration – for the detection of *Cryptosporidium* oocysts under ideal laboratory conditions and highly purified water. Since LW provided virtually no interference – there were nominally no other particles in the water except for the spiked *Cryptosporidium* – and the development of Optical Fingerprints required at least two species to be used, these measurements were generally performed with 2 μm PSL spheres as the second species (although some measurements were also made with 4 μm PSL spheres as well). 2 μm PSL spheres were chosen since while *Cryptosporidium* oocysts are roughly 4 μm – 5 μm in diameter, the scattered light intensities measured by the MALS Gen-4 instrument for *Cryptosporidium* were more similar to 2 μm PSL spheres.
- PSL spheres spiked into treated drinking water. These measurements tested the performance of the MALS system in a realistic water matrix using standard light scattering particles as targets that are symmetric, isotropic, uniform, and stable.
- *Cryptosporidium* oocysts spiked in drinking water. These measurements represented the goal of this project to demonstrate a proof-of-principle performance of a real-time, continuous detection system (MALS) with a realistic matrix – drinking water – using asymmetric and non-uniform biological targets.

The Relationship Between Experiments and Performance Parameters

The Performance Parameters discussed in Chapter 2 were related to the measurements performed above as follows:

- The Effective Patrol Volume Rate (EPVR), the measure of how much water is scanned per time by the instrument, was determined in purified laboratory water (LW). A LW sample was spiked, typically to a concentration of 250/mL or 500/mL, with a target particle such as a PSL sphere or *Cryptosporidium* oocyst. The instrument registered a certain number of events per time. The EPVR was calculated as follows:

$$EPVR = (\#events\ measured/time) / concentration \quad (3.1)$$

- The Identification Rate (IDR), a measure of the ratio of target particles correctly identified to the number that are registered as events, was determined in purified laboratory water (LW). The IDR was determined after an Optical Fingerprint was generated with either drinking water or other particles. If the other particles were PSL spheres, the IDR represented a best case for the specific instrument configuration and analysis parameters.
- The False Negative Rate (FNR), a measure of the fraction of target particles that pass through the system but were not correctly identified, was simply:

$$FNR = 1 - IDR \quad (3.2)$$

- The False Positive Rate (FPR), a measure of other particles, such as drinking water particles, that were incorrectly identified as the target particle (e.g., PSL spheres or *Cryptosporidium*), was determined in several ways depending upon the experiment performed. The FPR was measured as either a fraction of mis-identified particles, or as a number of mis-identified particles per volume. For the PSL spheres in LW, the FPR was determined from the number of non-target spheres that were incorrectly identified as the target sphere. This was also a form for species-species interference. For PSL spheres in drinking water, the FPR was determined from the number of drinking water particles that were incorrectly identified as the target sphere. For *Cryptosporidium* in LW, the FPR was determined as either an interference from PSL spheres (e.g. 2 µm PSL spheres) or as a number per volume based on events generated by LW itself. For *Cryptosporidium* in WPE, the FPR was determined from the number of drinking water particles that were incorrectly identified as *Cryptosporidium.*
- The Interference Factor (IF) was a measure of the decrease in the effective Identification Rate when target particles such as PSL spheres or *Cryptosporidium* were spiked into drinking water. Based on measurements in purified laboratory water, an Identification Rate (the ratio of correctly identified targets to total target events registered) was established. When the same concentration of targets was spiked into drinking water, the number of correctly identified particles generally decreased. The ratio of the number correctly identified (or presumed to be correctly identified, since there was no absolute method the verify that an identification in drinking water was correct) targets in drinking water to those correctly identified in purified laboratory water was the IF.

Polystyrene Latex Spheres in Purified Laboratory Water

Since the MALS system is a new technology, a significant number of measurements were performed using PSL microspheres, which are National Institute of Science and Technology (NIST) qualified light scattering standards, in order to establish baseline performance. Typically, spheres of size 1 μm, 2 μm, and 4 μm were measured. Since the PSL spheres are highly uniform and stable – unlike *Cryptosporidium* oocysts – they were useful for calibration of the MALS instrument and provided a means to measure system long-term stability.

The procedure used for measurement of PSL spheres in purified laboratory water (LW) was as follows (with exceptions noted in the discussion):

1. The flow rate through the system was set at 100 ± 5 mL/minute.
2. A LW sample was spiked with 250 spheres/mL or 500 spheres/mL of 1 μm PSL spheres and measurements using MALS were performed. Typically six to eight consecutive two minute observations were made. This was repeated for 2 μm PSL spheres and 4 μm PSL spheres.
3. The data were divided into two sets: a set used for training - the development of the optical fingerprint; and a set used for testing, to test the optical fingerprint and the performance of the system.
4. The performance of the system for PSL spheres in LW was measured using the following performance parameters (see Chapter 2 – Materials and Methods as well as the explanation above):
   - Effective Patrol Volume Rate (EPVR): the effective amount of water fully patrolled, or scanned per time for a given instrument configuration and target particle.
   - Identification Rate: The ratio of the number of target particles identified correctly to the number of target events detected in LW.
   - False Positive Rate: The ratio of the number of other targets identified incorrectly to the number of other targets detected in LW.

Results from three representative experiments using PSL spheres in LW are shown in Tables 3.1 through 3.3. The table format shown here is used through this Chapter. Since the instrument configurations (see "Instrument Configuration" in Chapter 2) were modified between experiments in order to improve performance, Tables 3.1 through 3.3 show the results from different instrument configurations, which included changes in detectors, amplifier electronics, laser beam profile, or analysis software. In these tables, "Particle Count" corresponds to measurements performed concurrently using the MET-One particle counter; "UNK" corresponds to the events that are unable to be identified with any of the three sphere species; "FP Rate," the False Positive Rate, is shown as "0.0%" if no False Positive events are measured but the maximum estimated False Positive Rate was determined by 1/(Number of Events Measured +1). "NM" means "not measured."

Table 3.1 shows the results from two minute measurements taken shortly after the Gen-4 instrument was completed. LW was spiked with a concentration of 500 spheres/mL for each species, although the MET-One particle count indicated that at least the 4 μm PSL spike may have been low. In this particular case, the water flow rate through the instrument was 10 mL/min, so the number of events collected was relatively low. These data therefore represent initial performance of the Gen-4 instrument. The Identification Rates for each of the sphere species was high: 92.1%, 96.9%, and 96.7%, for 1 μm, 2 μm, and 4 μm PSL spheres, respectively. The False Positive Rates (FPR) for all species was zero for all species, but the maximum estimated False Positive Rates were 0.22%, 0.26%, and 0.28% for 1 μm, 2 μm, and 4 μm PSL spheres, respectively. These relatively high estimated FPR were a consequence of the low number of events measured. The Effective Patrol Volume

Table 3.1
Measurements of initial performance of MALS instrument for PSL spheres in LW showing the results of two minute measurements for spheres spiked to 500/mL

		# of detected events identified as[3]:							
Target species[1]	Particle count[2]	1 µm PSL	2 µm PSL	4 µm PSL	"UNK"	ID Rate[4] (%)	FP Rate [5] (%)	EPVR[6] (mL/hr)	EVM[7] (mL)
Matrix[8] (purified water)	3.9								
1 µm PSL	NM	128	0	0	11	92.1	0.0 (<0.22)	8.3	0.28
2 µm PSL	NM	0	216	0	7	96.9	0.0 (<0.26)	13.4	0.45
4 µm PSL	359.5	0	0	232	8	96.7	0.0 (<0.28)	20.1	0.67

Notes:
1. This column shows the target species being measured.
2. This column shows the Met-One particle count, if performed. NM means "not measured." The Met-One works reliably for particles of diameter down to about 3 µm, so measurements of particles of smaller size are not shown.
3. The four possible outcomes for an event, in this case, are 1 µm PSL spheres, 2 µm PSL spheres, 4 µm PSL spheres, or "Unknown" (UNK).
4. This column shows the Identification Rate, that is, the number of correctly identified events to the total number of events measured.
5. This column shows the False Positive Rate, which in this case were misidentification of species, that is, the number of other PSL spheres incorrectly measured as the Target Species to the total number of other species events. The estimated maximum False Positive Rate is shown in parentheses.
6. EPVR is the Effective Patrol Volume Rate (in mL/hr) for each species, that is the total volume of water scanned per time for each species. The EPVR is calculated based on the assumed spiking concentration except for cases when direct particle counts were possible, such as the 4 µm case.
7. This column shows the Effective Volume Measured (in mL) by the system for the duration of the measurement. It is equal to the EPVR times the measurement time, in this case two minutes.
8. In the case of the matrix – purified laboratory water – typically 0 or 1 events are measured in 12 minutes of measurements.

Rates for each species were: 8.3 mL/hr, 13.4 mL/hr, and 20.1 mL/hr for 1 µm, 2 µm, and 4 µm PSL spheres, respectively.

Table 3.2 shows the results from a second set of measurements of PSL spheres each spiked to 500 spheres/mL in LW. These data were taken after the water flow rate through the instrument was increased to 100 mL/min. These results cover a larger sample than those shown in Table 3.1 – nearly 2000 events for each PSL species were collected over a two minute time. There was good discrimination between the species and no false positives were detected, so maximum estimated False Positive Rates are shown. The Identification Rates were 92.3%, 88.9% and 87.0% for 1 µm, 2 µm and 4 µm PSL spheres, respectively. The Effective Patrol Volume Rates for these measurements were: 108.9 mL/hr, 155.0 mL/hr, and 131.0 mL/hr, for 1 µm, 2 µm and 4 µm PSL spheres, respectively.

Table 3.2
Measurements of performance of MALS instrument for PSL spheres in LW. Flow rate of 100 mL/minute

Target species	Particle count	# of detected events identified as: 1 µm PSL	2 µm PSL	4 µm PSL	"UNK"	ID Rate (%)	FP Rate (%)	EPVR (mL/hr)	EVM (mL)
Matrix (purified water)	1.9								
1 µm PSL	NM	1580	0	0	131	92.3	0.0 (<0.021)	108.9	3.42
2 µm PSL	NM	0	2235	0	279	88.9	0.0 (<0.026)	155.0	5.03
4 µm PSL	446.9	0	0	1894	282	87.0	0.0 (<0.024)	131.0	4.87

Table 3.3
Third set of measurements of performance of MALS instrument for PSL spheres in LW. Measurements were six minutes long, generating more events.

Target species	Particle count	# of detected events identified as: 1 µm PSL	2 µm PSL	4 µm PSL	"UNK"	ID Rate (%)	FP Rate (%)	EPVR (mL/hr)	EVM (mL)
Matrix (purified water)	NM								
1 µm PSL	NM	4740	0	0	73	98.5	0.0 (<0.007)	96.3	9.63
2 µm PSL	NM	0	6870	1	45	99.3	0.0 (<0.008)	138.3	13.83
4 µm PSL	462.5	0	0	6741	24	99.6	0.009	146.3	14.63

Table 3.3 shows the results from a third set of measurements of PSL spheres in LW. The spiking concentration for each target species was 500 spheres/mL and the measurements were six minutes each. The water flow rate through the instrument was 100 mL/min. These results are consistent with the previous results shown in Tables 3.1 and 3.2 with a larger sample than those shown in Table 3.2. More than 4,700 1 µm PSL sphere events, more than 6,800 2 µm PSL sphere events and more than 6,700 4 µm PSL sphere events were collected. There is high discrimination between the spheres and only a single false positive occurred when a 2 µm PSL sphere was incorrectly identified as a 4 µm PSL sphere. For the 2 µm PSL sphere this corresponded to a False Positive Rate of 0.009%. The maximum estimated False Positive Rates for 1 µm PSL

spheres and 2 µm PSL spheres were 0.007% and 0.008%, respectively. The Identification Rates were 98.5%, 99.3% and 99.6% for 1 µm, 2 µm and 4 µm PSL spheres, respectively. The Estimated Patrol Volume Rates for these measurements were 96.3 mL/min, 138.3 mL/min, and 146.3 mL/min for 1 µm, 2 µm and 4 µm PSL spheres, respectively.

Tables 3.1 through 3.3 illustrate the consistent and high performance of the MALS system for discrimination of PSL spheres in LW. PSL spheres of significantly different diameters were relatively easy to discriminate since they are symmetric and uniform, and generated very low False Positive Rates in purified laboratory water. Discrimination of *Cryptosporidium* was more challenging because of the asymmetric shape and non-uniformity of the oocysts as will be seen in the following section.

Cryptosporidium in Purified Laboratory Water: Determining the False Positive Rate in Purified Laboratory Water vs. PSL Spheres

Results from two representative experiments of *Cryptosporidium* spiked into purified laboratory water are shown in Tables 3.4 and 3.5. Each table represents results from different instrument configurations, which included changes in detectors, amplifier electronics, laser beam profile, or analysis software as well as different *Cryptosporidium* oocyst batches. Analysis was not performed to determine batch-to-batch oocyst consistency. In Tables 3.4 and 3.5, "Particle Count" corresponds to measurements performed concurrently using the MET-One particle counter; "UNK" corresponds to the events that were unable to be identified with any of the target species; "FP Rate," the False Positive Rate was determined by PSL sphere species incorrectly identified as *Cryptosporidium*, shown as "0.0%" if no False Positive events were measured but the maximum estimated False Positive Rate was determined by 1/(Number of Events Measured +1); "NM" means "not measured."

Table 3.4 shows the results from the first measurements of the Gen-4 MALS instrument of *Cryptosporidium* oocysts in laboratory water. These were the first experiments using oocysts purified using flow cytometry cell sorting (FCCS) by the Wisconsin State Laboratory of Hygiene (WSLH). The water flow rate through the instrument was 100 mL/min. An optical fingerprint was developed using a spike of 2 µm PSL spheres at 500 spheres/mL into LW, 4 µm PSL spheres at 500 spheres/mL into LW and *Cryptosporidium* at 250 oocysts/mL into LW. The Identification Rate for *Cryptosporidium* with these interference species was 47.7%. No False Positives were measured, so the measured False Positive Rate was 0.0%, but was estimated to be a maximum of 1/(Number of 2 µm events + number of 4 µm events + 1) or 0.027%. The Effective Patrol Volume Rate for *Cryptosporidium* was 88.6 mL/hr.

Table 3.5 shows the results from a second experiment using the Gen-4 MALS instrument of *Cryptosporidium* oocysts in laboratory water. In this case, only 2 µm PSL spheres were used as the species interference. An optical fingerprint was developed using a spike of 2 µm PSL spheres at 250 spheres/mL into LW, and *Cryptosporidium* at 250 oocysts/mL into LW. The Identification Rate for *Cryptosporidium* with this interferences was 63.9%, and the False Positive Rate (against 2 µm spheres) for this measurement was 0.28%. The Effective Patrol Volume Rate for *Cryptosporidium* in this case was 10.8 mL/hr, substantially lower than the results shown in Table 3.4. The analytical variables were adjusted to increase the Identification Rate. However, it is important to note that the higher Identification Rate was achieved in this case, 63.9%, versus the results shown in Table 3.4 (47.7%), at the expense of a higher False Positive Rate (0.28% here vs. 0.0% in Table 3.4) and lower Effective Patrol Volume Rate.

Table 3.4
Measurements of initial performance of MALS instrument for identifying *Cryptosporidium* oocysts in LW using 2 µm and 4 µm PSL spheres as interference

Target species	Particle count	# of detected events identified as: 2 µm PSL	4 µm PSL	*Crypto*	"UNK"	ID Rate (%)	FP Rate (%)	EPVR (mL/hr)	EVM (mL)
Matrix (purified water)	1.9								
2 µm PSL[1]	NM	1347	0	0	157	89.6	0.0 (<0.026)	338.3	3.01
4 µm PSL[2]	446.9	0	1742	0	453	79.4	0.0 (<0.031)	395.2	4.91
Crypto[3]	NM	0	0	805	882	47.7	0.0 (<0.027)	88.6	6.75

Notes:
1. 32 second measurement
2. 45 second measurement
3. 4.5 minute measurement

Table 3.5
Measurements of performance of MALS instrument for identifying *Cryptosporidium* oocysts in LW using 2 µm PSL spheres as the only interference

Target species	Particle count	# of detected events identified as: 2 µm PSL	*Crypto*	"UNK"	ID Rate (%)	FP Rate (%)	EPVR (mL/hr)	EVM (mL)
Matrix (purified water)	NM							
2 µm PSL[1]	NM	1211	5	565	68.00	0.0 (<0.22)	35.62	7.12
Crypto[2]	NM	0	287	162	63.92	0.28	10.80	1.80

Notes:
1. 12 minute measurement
2. 10 minute measurement

Polystyrene Latex Spheres in Drinking Water (WPE): Determining the Interference Factor for Ideal Particles in Finished Drinking Water Samples

The measurement of spiked samples of PSL spheres in finished drinking water (in this case, Weymouth Plant Effluent, or "WPE") provided a means to test the MALS instrument in realistic drinking water with standard particles that did not have the morphological complexity or variability of *Cryptosporidium* oocysts. The PSL spheres were highly stable, symmetric, and consistent, while the constituents of the WPE matrix were generally unknown and variable. Thus, these measurements provided a next step towards direct measurement of *Cryptosporidium* in drinking water.

Results from three different test runs, corresponding to three different Gen-4 MALS configurations, are shown in the following section. The most important performance parameters for these measurements was the Interference Factor, which measured the level of interference generated by the background matrix water. The Interference Factor was by definition 0% in purified laboratory water and 100% in waters where the particle density generated continuous particle events in the laser beam and effectively "blinded" the system.

Test Run #1

Table 3.6 and Figure 3.1 show results from spiking experiments with 2 μm and 4 μm PSL spheres in LW and WPE. The particle count for the WPE matrix #0202-1 (for particles of diameter > 2 μm) was 65.1 particles/mL. An optical fingerprint was generated using LW spiked with 2 μm and 4 μm PSL spheres at 500 spheres/mL, respectively, and unspiked WPE. The Identification Rate, the False Positive Rate and the Effective Patrol Volume Rate (EPVR) were determined, as earlier. The Identification Rates for 2 μm and 4 μm spheres were 42.0% and 25.9%, respectively. The False Positive Rates were 0.0% for both 2 μm and 4 μm spheres (no particles in the WPE were incorrectly identified as PSL spheres), with an estimated maximum (based on the number of WPE events) of 0.058%. The ID Rate was significantly lower than the ID Rates seen in LW, which was expected due to the interfering WPE matrix. The Interference Factors – the loss in expected counts per volume due to the WPE matrix – for 2 μm and 4 μm PSL spheres spiked into the WPE were 4.5% and 0.8%, respectively. It should be noted that the LW samples, in this case, were spiked at 500 spheres/mL and the WPE samples were spiked at 300 spheres/mL, so the Interference Factor is adjusted to correspond to equivalent concentration. Figure 3.1 shows these results graphically. The columns on the left are the mean number of (equivalent) PSL sphere events per time (2 minutes) for the LW spiked samples, the columns in the center are the mean number of false positive events from the WPE sample (in this case none), and the columns on the right are the mean number of (equivalent) PSL sphere events per time for the WPE spiked samples. Three samples were used for the LW results and six were used for the WPE results. The error bars reflect one standard deviation around the mean.

Test Run #2

A second set of measurements were performed using various spiking concentrations of 1 μm, 2 μm, and 4 μm PSL spheres individually in a WPE matrix with a different configuration of the MALS system.

Table 3.6
Measurements of performance of the MALS instrument for identification of 2 µm and 4 µm PSL spheres in WPE

Target species/ matrix	PC/ *spike*[1]	Total # of detected events identified as: WPE	2 µm PSL	4 µm PSL	UNK	ID Rate (%)	FP Rate[2] (%)	EPVR (mL/hr)[3]	EVM (mL)[3]	Interfer. Factor[4] (%)
Matrix[5] (WPE)	65.1	559	0	0	1163	[6]	[7]	*132.25*	*26.45*	
2 µm PSL/LW[8]	*500/mL*	18	996	0	1357	42.0	0.0 (< 0.058)	79.0	4.74	0
4 µm PSL/LW[8]	*500/mL*	37	0	1631	4626	25.9	0.0 (< 0.058)	209.8	12.59	0
2 µm PSL/WPE[5]	*300/mL*	280	1903	0	3166	[6]	[7]	[9]	[9]	4.5
4 µm PSL/WPE[5]	*300/mL*	1605	0	3235	10586	[6]	[7]	[9]	[9]	0.8

Notes:
1. Normal type indicates a particle count (PC) measurement of the sample on the MET-One particle counter, and italics indicates that no PC measurement was made and the number indicates the target (estimated) spiking concentration.
2. Percentage of WPE events incorrectly identified as the specified species of PSL spheres.
3. The Effective Patrol Volume Rate and Effective Volume Measured were estimated for the WPE, since an actual particle count was not possible with the Met-One particle counter – below 3 µm, particles in the water are not accurately counted. Estimates are shown in italics.
4. The Interference Factor was the decrease in counts/time of a target particle when it is measured directly in an interfering matrix such as WPE, as compared to the counts/time when measured in purified laboratory water for identical spiking concentrations. Therefore it is defined as 0% in LW.
5. Six 2-minute measurements.
6. The Identification Rate was determined in purified laboratory water only.
7. The False Positive Rate was determined from unspiked WPE only.
8. Three 1.2-minute measurements.
9. The Effective Patrol Volume Rate and the Effective Volume Measured were difficult to estimate for a spike into an uncharacterized matrix.

Table 3.7 and Figure 3.2 show the results for 4 µm PSL spheres spiked into WPE. The Met-One particle count (for particles of size > 2 µm) for the WPE matrix sample was 117.0 particles/mL. An Optical Fingerprint was generated using LW spiked with 4 µm PSL spheres at 250 spheres/mL and WPE. The Identification Rate for this case was 29.0%, with a low False Positive Rate of 0.03% (only one event out of 2972 WPE events). The Effective Patrol Volume Rate was high, 342.24 mL/hr, and the Interference Factor was 26.0%. Five two-minute samples were gathered for WPE and six for the 4 µm sphere spiked samples (LW and WPE), however, due to an anomaly in data analysis, only 112 seconds of the 120 seconds were used. Figure 3.2 shows these results graphically. The left bar is the mean number of (equivalent) PSL sphere events

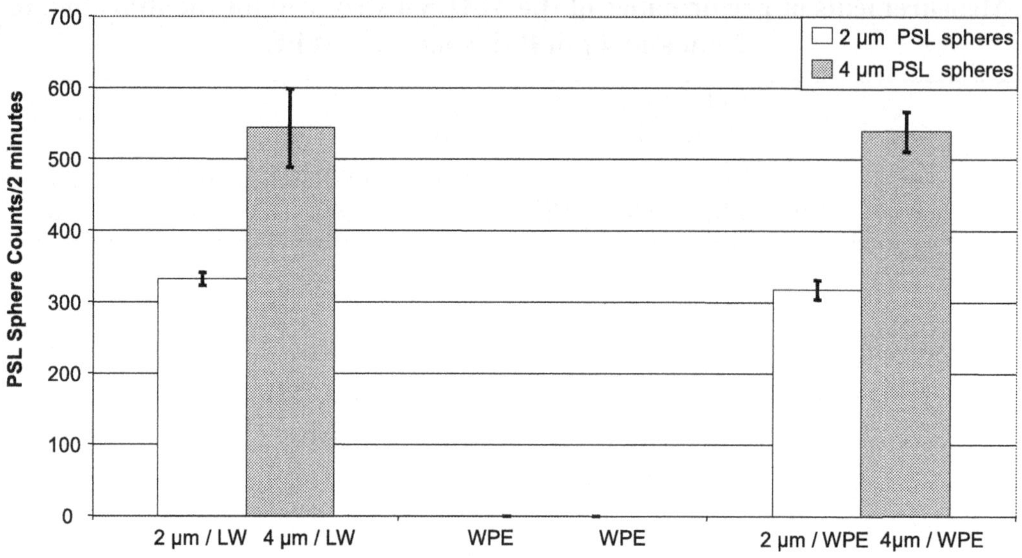

Figure 3.1 Interference Factor for 2 µm and 4 µm PSL spheres in WPE for Test Run #1. The Interference Factor is measured as 4.5% for 2 µm spheres and 0.8% for 4 µm spheres.

Table 3.7
Measurement of the performance of the MALS instrument for 4 µm PSL spheres spiked at 250/mL in WPE

Target species/ matrix	PC/ *spike*[1]	Total # of detected events identified as: WPE	4 µm PSL	UNK	ID Rate (%)	FP Rate[2] (%)	EPVR (mL/hr)[3]	EVM (mL)[3]	Interfer. Factor (%)
Matrix[4] (WPE)	117.0	993	1	1978			*163.30*	*25.40*	
4 µm PSL/LW[5]	256.3/mL	137	4623	11210	29.0	0.03	333.80	62.31	0
4 µm PSL/WPE[5]	340.0/mL	954	3422	9311			*215.66*	*40.26*	26.0

Notes:
1. Normal type indicates a particle count (PC) measurement of the sample on the MET-One particle counter, and italics indicates that no PC measurement was made and the number indicates the target spiking concentration.
2. The False Positive Rate is determined from unspiked WPE only.
3. Estimated values are in italics.
4. Five 112 second measurements.
5. Six 112 second measurements.

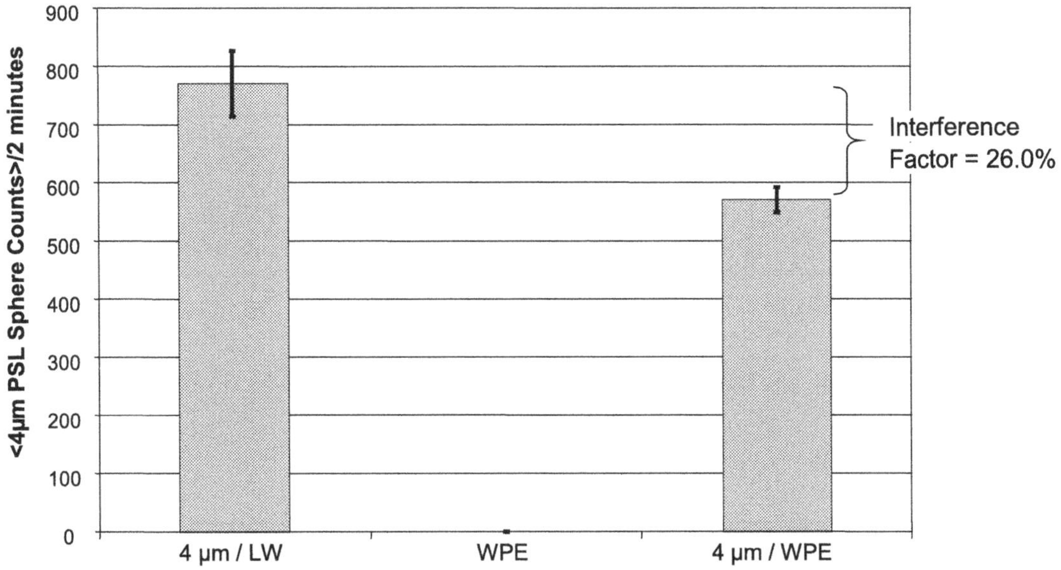

Figure 3.2 Interference Factor for 4 µm PSL spheres in WPE. The Interference Factor was 26.0% for 4 µm spheres spiked at a concentration of 250 spheres/mL for this experiment.

per time (112 seconds) for the LW spiked samples, the center bar is the mean number of false positive events from the WPE sample, and the right bar is the mean number of (equivalent) PSL sphere events per time for the WPE spiked samples. Six samples were used for the 4 µm/LW and 4 µm/WPE results and five were used for the WPE results. The error bars reflect one standard deviation around the mean.

Table 3.8 and Figure 3.3 show the results from a measurement of 1 µm PSL spheres spiked into WPE. The Met-One particle count (for particles of size > 2 µm) for the WPE matrix sample was 46.7 particles/mL. An optical fingerprint was generated using LW spiked with 1 µm PSL spheres at 500 spheres/mL, respectively and WPE. The Identification Rate for this case was 32.3%, with a False Positive Rate of 0.12%. The Effective Patrol Volume Rate for the 1 µm PSL spheres was 55.67 mL/hr. The Interference Factor was 90.9%, indicating a high interference from the WPE water. This was much higher than previous measurements of 2 µm PSL spheres and 4 µm PSL spheres in WPE. Six two minute samples were gathered for all cases, however, due to an anomaly in data analysis, only 112 seconds of the 120 seconds were used. Figure 3.3 shows these results graphically. The left bar is the mean number of (equivalent) PSL sphere events per time (112 seconds) for the LW spiked samples, the center bar is the mean number of false positive events from the WPE sample, and the right bar is the mean number of (equivalent) PSL sphere events per time for the WPE spiked samples. Six samples were used for all cases. The error bars reflect one standard deviation around the mean.

Table 3.9 and Figure 3.4 show the results from a measurement of 2 µm PSL spheres spiked into WPE. The MET-One particle count (for particles of size > 2 µm) for this WPE matrix sample was 54.9 particles/mL. An optical fingerprint was generated using LW spiked with 2 µm PSL spheres at 1000 spheres/mL, respectively and WPE. The Identification Rate for this case was 30.5%, with a False Positive Rate of 0.04%. The Effective Patrol Volume Rate for 2 µm PSL spheres was 172.85 mL/hr and the Interference Factor was 33.4%. Five two minute samples were gathered for

Table 3.8
Measurement of the performance of the MALS instrument for 1 μm PSL spheres at 500/mL in WPE

Target species/ matrix	Total # of detected events identified as: PC/ *spike*[1]	WPE	1 μm PSL	UNK	ID Rate (%)	FP Rate (%)	EPVR (mL/hr)	EVM (mL)	Interfer. Factor (%)
Matrix[2] (WPE)	46.7	3414	7	2252			*650.77*	*121.48*	
1 μm PSL/LW[2]	*500/mL*	54	1678	3464	32.3	0.12	55.67	10.39	0
1 μm PSL/WPE[2]	*500/mL*	3649	152	3734					90.9

Notes:
1. Normal type indicates a particle count (PC) measurement of the sample on the MET-One particle counter, and italics indicates that no PC measurement was made and the number indicates the target/estimated spiking concentration.
2. Six 112 second measurements.

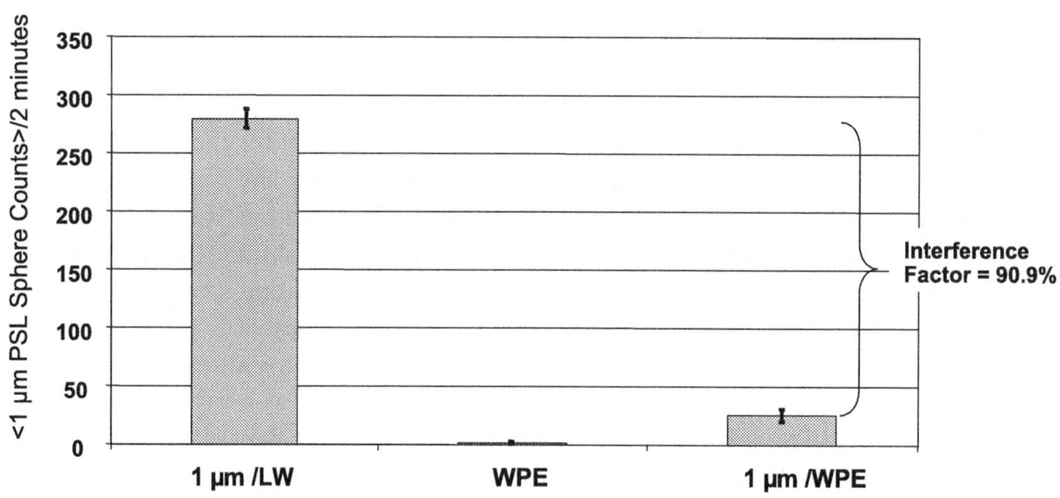

Figure 3.3 Interference Factor for 1 μm PSL spheres in WPE. The Interference Factor was 90.9% for 1 μm spheres spiked into WPE at a concentration of 500 spheres/mL.

Table 3.9
Measurement of the performance of the MALS instrument for 2 μm PSL spheres at 1000/mL in WPE

Target species/ matrix	PC/ *spike*[1]	Total # of detected events identified as: WPE	2 μm PSL	UNK	ID Rate (%)	FP Rate (%)	EPVR (mL/hr)	EVM (mL)	Interfer. Factor (%)
Matrix[2] (WPE)	54.9/mL	2350	2	2365			*552.34*	*85.92*	
2 μm PSL/LW[3]	*1000/mL*	258	9829	22179	30.5	0.04	172.85	32.27	0
2 μm PSL/WPE[3]	*1000/mL*	1957	6544	16921					33.4

Notes:
1. Normal type indicates a particle count (PC) measurement of the sample on the MET-One particle counter, and italics indicates that no PC measurement was made and the number indicates the target spiking concentration.
2. Five 112 second measurements.
3. Six 112 second measurements.

WPE and six two minute samples for the 2 μm sphere spiked samples (LW and WPE), however, due to an anomaly in data analysis, only 112 seconds of the 120 seconds were used. Figure 3.4 shows these results graphically. The left bar in Figure 3.4 is the mean number of (equivalent) PSL sphere events per time (112 seconds) for the LW spiked samples, the center bar is the mean number of false positive events from the WPE sample, and the right bar is the mean number of (equivalent) PSL sphere events per time for the WPE spiked samples. Six samples were used for the 2 μm/LW and 2 μm/WPE results and five were used for the WPE results. The error bars reflect one standard deviation around the mean.

Test Run #3

A third set of measurements were performed using 2 μm and 4 μm PSL spheres spiked individually in a WPE matrix with another configuration of the MALS system.

Table 3.10 and Figure 3.5 show the results from measurements of 4 μm PSL spheres spiked into WPE. The MET-One particle count (for particles of size > 2 μm) for this WPE matrix sample was 126.0 particles/mL. An two-species optical fingerprint was generated using LW spiked with 4 μm PSL spheres at 250 spheres/mL, respectively and WPE. The Identification Rate for this case was 68.6%, with a False Positive Rate of 0.79%. The Effective Patrol Volume rate for the 4 μm PSL spheres is 73.5 mL/hr and the Interference Factor is 35.6%. Six two minute samples were gathered for all cases. The left bar in Figure 3.5 is the mean number of PSL sphere events per time for the LW spiked samples, the center bar is the mean number of false positive events from the WPE sample, and the right bar is the mean number of PSL sphere events per time for the WPE spiked samples. Six samples were used for the 4 μm/LW and 4 μm/WPE results and five were used for the WPE results. The error bars reflect one standard deviation around the mean.

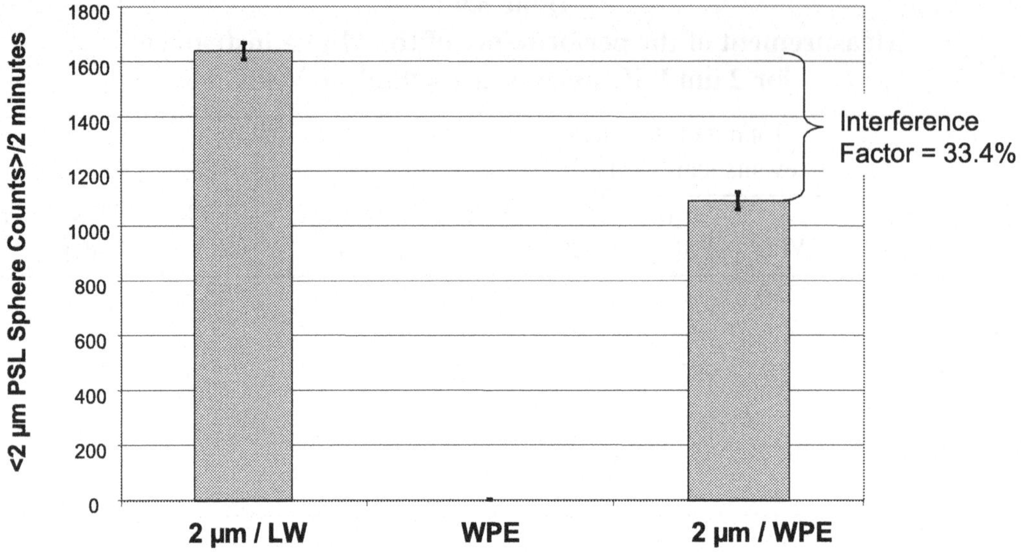

Figure 3.4 Interference Factor for 2 µm PSL spheres in WPE. The Interference Factor was 33.4% for 2 µm spheres spiked at a concentration of 1000 spheres/mL.

Table 3.10
Measurement of the performance of the MALS instrument for 4 µm PSL spheres at 250/mL in WPE

Target species/ matrix	PC/ *spike*[1]	Total # of detected events identified as: WPE	4 µm PSL	UNK	ID Rate (%)	FP Rate (%)	EPVR (mL/hr)	EVM (mL)	Interfer. Factor (%)
Matrix[2] (WPE)	126.0	498	9	638			*45.45*	*9.09*	
4 µm PSL/LW[2]	*250/mL*	13	2412	1003	70.4	0.77	68.60	13.72	0
4 µm PSL/WPE[2]	*250/mL*	373	1553	1154					35.6

Notes:
1. Normal type indicates a particle count (PC) measurement of the sample on the MET-One particle counter, and italics indicates that no PC measurement was made and the number indicates the target spiking concentration.
2. Six 120 second measurements.

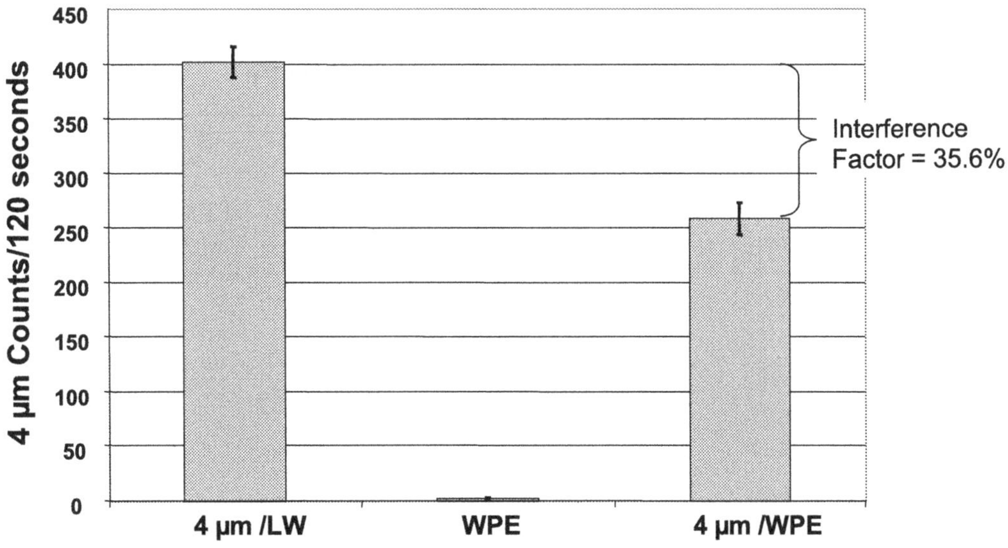

Figure 3.5 Interference Factor for 4 µm PSL spheres in WPE, based on data collected on the MALS system. The Interference Factor was 35.6% for 4 µm spheres spiked at a concentration of 250 spheres/mL.

Table 3.11 and Figure 3.6 show the results from measurements of 2 µm PSL spheres spiked into WPE. The MET-One particle count (for particles of size > 2 µm) for this WPE matrix sample was 94.6 particles/mL. A two-species optical fingerprint was generated using LW spiked with 2 µm PSL spheres at 500 spheres/mL and WPE. The Identification Rate for this case was 87.4%, with a False Positive Rate of 0.18%. The Effective Patrol Volume Rate was 75.58 mL/hr for the 2 µm PSL spheres and the Interference Factor was 79.1%. Six two minute samples were gathered for all cases. The left bar in Figure 3.6 is the mean number of PSL sphere events per time for the LW spiked samples, the center bar is the mean number of false positive events from the WPE sample, and the right bar is the mean number of PSL sphere events per time for the WPE spiked samples. Six samples were used for all cases. The error bars reflect one standard deviation around the mean.

Table 3.12 is a summary of the results for the PSL spheres spiked and identified in drinking water (WPE). Figure 3.7 shows the Interference Factor as a function of PSL sphere diameter for the three different test sets corresponding to three different configurations of the MALS system.

Several observations can be made about these results:

- Each configuration of the MALS instrument (e.g., laser profile, electronics, detector set, and analysis software) generated different results. This is an expected result. In general, results from one configuration cannot be directly compared to results from a different configuration.

Table 3.11
Measurement of the performance of the MALS instrument for 2 µm PSL spheres at 500/mL in WPE

Target species/ matrix	PC/ *spike*[1]	Total # of detected events identified as: WPE	2 µm PSL	UNK	ID Rate (%)	FP Rate (%)	EPVR (mL/hr)	EVM (mL)	Interfer. Factor (%)
Matrix[2] (WPE)	94.6/mL	1635	4	586			*117.07*	*23.41*	
2 µm PSL/LW[2]	*500/mL*	135	6608	815	87.4	0.18	75.58	15.12	0
2 µm PSL/WPE[2]	*500/mL*	1467	1380	2400					79.1

Notes:
1. Normal type indicates a particle count (PC) measurement of the sample on the MET-One particle counter, and italics indicates that no PC measurement was made and the number indicates the target spiking concentration.
2. Six 120 second measurements.

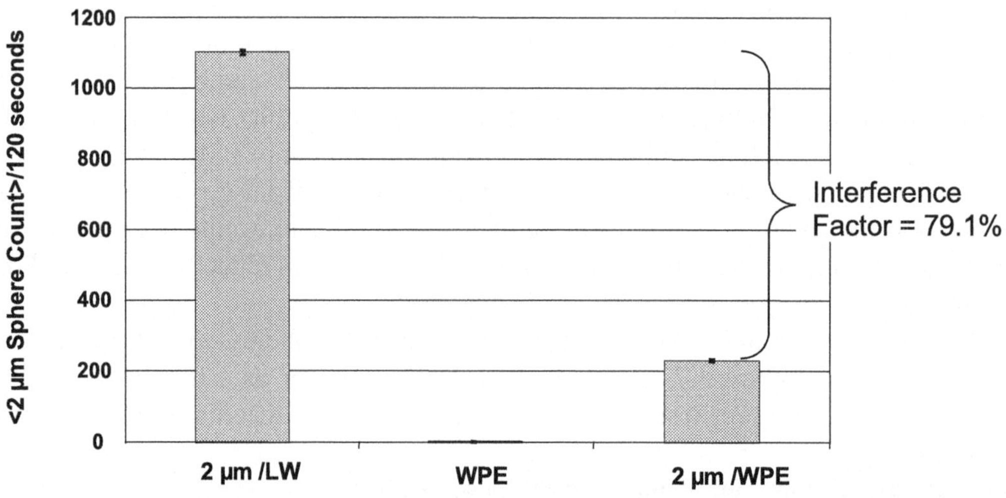

Figure 3.6 Interference Factor for 2 µm PSL spheres in WPE. The Interference Factor was 79.1% for 2 µm spheres spiked at a concentration of 500 spheres/mL.

Table 3.12
Summary of PSL sphere spiking results in WPE, arranged by size and test run number

Species	Test run[1]	[Spike] [2]	WPE PC[3]	ID Rate[4] (%)	FP Rate[5] (%)	EPVR[6]	IF[7] (%)
1 µm PSL	#2	500	46.7	32.3	0.12	55.67	90.9
2 µm PSL	#1	500/300	65.1	42.0	0.0 (< 0.058)	79.0	4.5
2 µm PSL	#2	1000	54.9	30.5	0.04	172.85	33.4
2 µm PSL	#3	500	96.4	87.4	0.18	75.58	79.1
4 µm PSL	#1	500/300	65.1	25.9	0.0 (< 0.058)	209.8	0.8
4 µm PSL	#2	250	117.0	29.0	0.03	333.80	26.0
4 µm PSL	#3	250	126.0	70.4	0.77	68.60	35.6

Notes:
1. Each test run corresponds to a different MALS configuration.
2. Target/estimated spiking concentration, in number of spheres per mL. If two numbers are shown, the first is the target spiking concentration in LW and the second was the target spiking concentration in WPE.
3. Particle count (> 2 µm) for the WPE matrix used for that specific spiking experiment.
4. Identification Rate.
5. False Positive Rate. The ratio of number of particles incorrectly identified as PSL spheres in a measurement of blank (unspiked) WPE matrix.
6. Estimated Patrol Volume Rate in mL/hr.
7. Interference Factor.

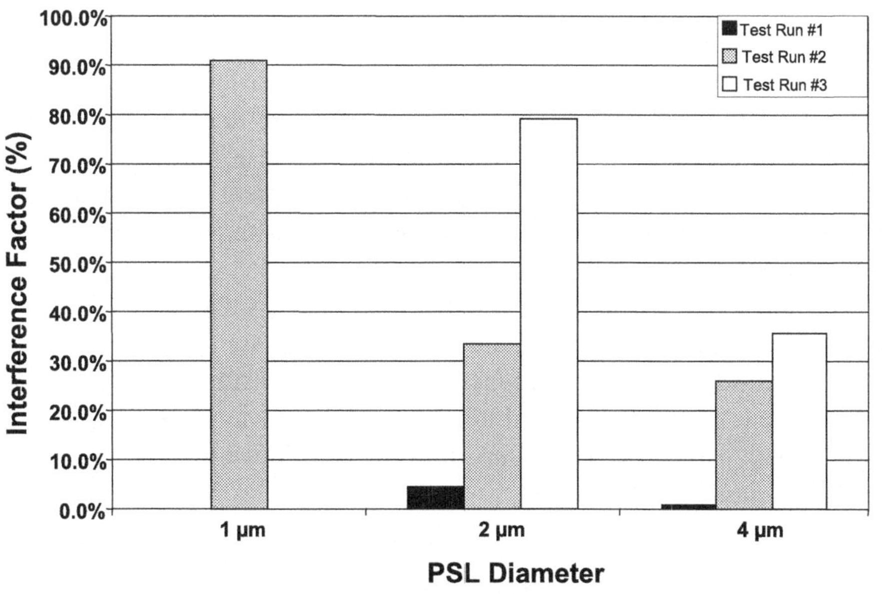

Figure 3.7 Summary of Interference Factor of PSL spheres in WPE as a function of test run and PSL sphere diameter

- While PSL spheres are a useful tool to evaluate performance, the goal of this project was to measure the performance of MALS for the detection of *Cryptosporidium* in drinking water, and as such, efforts were aimed towards improving performance relative to *Cryptosporidium*, not test spheres. As will be seen in following sections, despite good performance for PSL spheres in WPE (e.g., relatively low Interference Factors, except for 1 µm spheres, acceptable Identification Rates, and relatively low false Positive Rates and the general conclusion that PSL spheres are qualitatively distinguishable from other particles typically found in these water samples), detection of *Cryptosporidium* proved more challenging in drinking water, most likely due to the aspherical nature of *Cryptosporidium* oocysts and the inherent variability from individual oocyst to oocyst.
- The matrix water (WPE) was a significant variable. This was shown in the particle count results, which varied from 46.7 particles/mL to 126.0 particles/mL for these data. It is obvious that for very high particulate raw waters, the Colorado River Water (CRW) or the State Project Water (SPW), the MALS configurations used for finished drinking water here failed. The Interference Factor was 100% in these waters. In contrast, in purified laboratory water (effectively particle-free), the Interference Factor was 0%. The Interference Factor was clearly a function of water quality, most specifically particulate content, but this relationship was not investigated further in the current project. It is also clear that the MET-One particle counts were not sensitive enough to be used to determine MALS interference, since the MET-One particle counter only detected particles larger than 2 µm, but MALS detected and identified particles that were smaller, at least as small as 600 nm (PointSource, 2002).
- These experiments were performed to understand the general behavior of the system using idealized particles, so the spiking concentrations were not always held constant. Spiking concentrations between 250 spheres/mL and 1000 spheres/mL were used. While this range was well within the MALS ability to detect particles, there may have been effects of concentration on the Interference Factor, including self-interference, in this concentration range.
- The features in software used to determine if a particle has passed through the laser beam and the optical fingerprint parameters were variables that may be set to different values, depending upon the priorities such as high Identification Rate or low False Positive Rate. For example, using a more selective feature set which yields a higher Identification Rate generally coming at the cost of a higher False Positive Rate and lower Effective Patrol Volume Rate. This is consistent with behaviors of the receiver operator characteristics (ROC) curves that are well-known in the detection theory field (c.f., Macmillan and Creelman, 1991).
- One conclusion that can be drawn from Figure 3.7, is that for a specific test run, during which the instrument configuration and analysis parameters were fixed, the Interference Factor appeared to increase with decreasing PSL sphere diameter. In each of the three configurations, the Interference Factor was monotonically decreasing with particle size. This observation has importance relative to the *Cryptosporidium* results that follow in the next section.

Cryptosporidium in Drinking Water (WPE, OCGW): Determining the Interference Factor for *Cryptosporidium* Oocysts in Finished Drinking Water Samples

As was discussed in Chapter 2, Materials and Methods, the 3rd generation MALS instrument demonstrated unacceptable day-to-day instability. Disinfection required disassembly and exposure of sensitive parts to aggressive oxidizing agents such as H_2O_2. Once this problem was resolved with the implementation of the 4th generation MALS system, for which all of the sensitive optical components were no longer in direct contact with the flow stream, another problem arose. It was found that poor sample quality due to excessive concentration of bacterial and detritus contamination generated poor quality Optical Fingerprints. This problem was solved by the use of flow cytometry separated oocysts that generated much cleaner MALS optical fingerprints.

Initial experiments using the Gen-4 MALS system indicated that the Interference Factors for *Cryptosporidium* oocysts in finished drinking water were about 80%, which were unsatisfactorily high, considering similar measurements using PSL spheres in drinking water. It appeared in spikes of finished drinking water that the concentration was about ⅕ of the expected concentration. There were several possible explanations for this: (1) The initial experiments were anomalous, giving wrong results; (2) Something in the experimental procedure caused the loss; (3) Additional *Cryptosporidium* was added to water and the analytical techniques were unable to detect higher concentrations due to particle overlap or something related – i.e., it was concentration sensitive; (4) The water chemistry of the laboratory water in which the *Cryptosporidium* was measured for the optical fingerprint and the water chemistry of the drinking water in which *Cryptosporidium* was spiked was sufficiently different to cause a physical change in the oocyst that was manifested in the optical signature; (5) The *Cryptosporidium* clung to the particles in drinking water, thereby changing the optical fingerprint and making the oocysts part of the background; (6) Some other, unknown measurement and/or analytical problem; (7) A combination of some or all of the above.

Therefore, a series of experiments were executed to test whether the losses: were an experimental anomaly; were related to spiking concentration; were due to water chemistry differences between the LW and the drinking water matrices; were due to losses due to sticking of the oocysts to test vessels and tubing; or were due to analytical or electronic problems with the Gen-4 system. As will be shown in the following results, the anomalous loss of oocysts was solved essentially through software and electronics.

The procedure used for these experiments was the same as used for PSL spheres spiked into WPE.

Initial Measurements of Cryptosporidium in Drinking Water

Table 3.13 and Figure 3.8 show the results from measurements of *Cryptosporidium* spiked into WPE. The MET-One particle count (for particles of size > 2 μm) for this WPE matrix sample was 70.1 particles/mL. A two-species Optical Fingerprint was generated using LW spiked with *Cryptosporidium* at 250 spheres/mL and blank WPE. The Identification Rate was 1.4%, with a False Positive Rate of 0.044%. Effective Patrol Volume Rate was 213.5 mL/hr and the Interference Factor was 86.9%. Six two minute samples were gathered for all cases. The left bar in Figure 3.19 is the mean number of *Cryptosporidium* events per time for the LW spiked samples, the center bar is the mean number of false positive events from the WPE sample, and the right bar

Table 3.13
Measurement of the performance of the MALS instrument for *Cryptosporidium* oocysts spiked at 250/mL in WPE

Target species/ matrix	PC/ *spike*[1]	Total # of detected events identified as: WPE	*Crypto*	UNK	ID Rate (%)	FP Rate (%)[2]	EPVR (mL/hr)	EVM (mL)	Interfer. Factor (%)
Matrix[3] (Blank WPE)	70.1	1333	5	11483		0.044	*914.48*	*182.90*	
Crypto/LW[3]	*250/mL*	34	153	10488	1.4		213.50	42.70	0
Crypto/WPE[3]	*250/mL*	287	20	8969					86.9

Notes:
1. Normal type indicates a particle count (PC) measurement of the sample on the MET-One particle counter, and italics indicates that no PC measurement was made and the number indicates the target spiking concentration.
2. For the two-species *Cryptosporidium* results in this table and the tables that follow, the False Positive Rate will be shown in the same row as the Blank WPE, since the Blank WPE generates the false positives.
3. Six 120 second measurements.

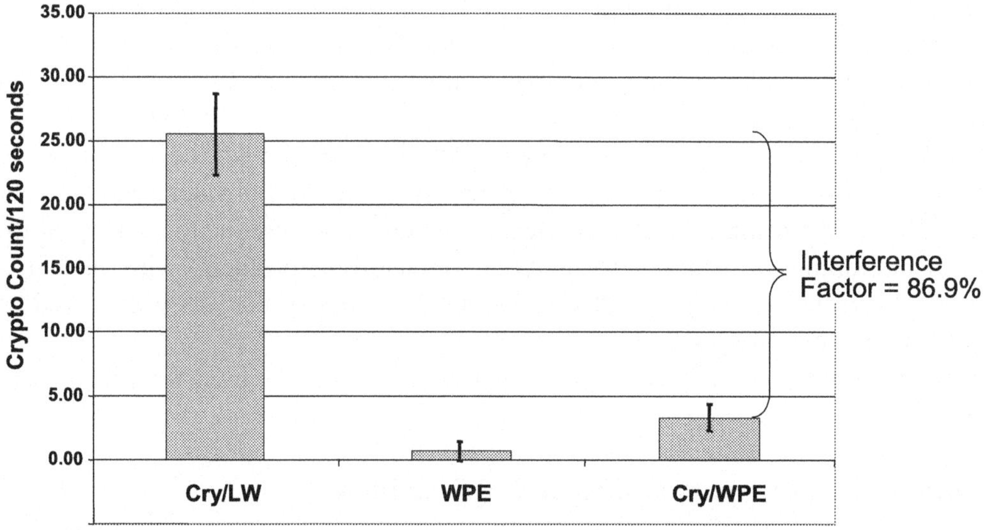

Figure 3.8 Interference Factor for *Cryptosporidium* oocysts in WPE. The Interference Factor was 86.9% for *Cryptosporidium* oocysts spiked at a concentration of 250 oocysts/mL.

Table 3.14
Measurement of the performance of the MALS instrument for *Cryptosporidium* oocysts spiked to 250/mL in Orange County Ground Water (OCGW)

Target species/ matrix	PC/ *spike*[1]	Total # of detected events identified as: OCGW	*Crypto*	UNK	ID Rate (%)	FP Rate (%)	EPVR (mL/hr)	EVM (mL)	Interf. Factor (%)
Matrix[2] (OCGW)	389.3	870	1	2798		0.027	*47.12*	*9.42*	
Crypto/LW[2]	*250/mL*	4	2747	9044	23.3		235.9	47.18	0
Crypto/OCGW[2]	*250/mL*	711	0	1968					100

Notes:
1. Normal type indicates a particle count (PC) measurement of the sample on the MET-One particle counter, and italics indicates that no PC measurement was made and the number indicates the target spiking concentration.
2. Six 120 second measurements.
"NA" - Not applicable.

is the mean number of *Cryptosporidium* events per time for the WPE spiked samples. Six samples were used for all cases. The error bars reflect one standard deviation around the mean.

These results show a very low Identification Rate which was somewhat mitigated by the high Effective Patrol Volume Rate. The False Positive Rate was low. For this case it corresponded to approximately five false identifications every 182.9 mL, or about 37 false identifications per liter of drinking water. However, the Interference Factor is very high: roughly 87% of the *Cryptosporidium* expected to be detected in WPE were not detected.

Measurements of Cryptosporidium in Drinking Water – OCGW Matrix

Table 3.14 shows the results from measurements of *Cryptosporidium* spiked into Orange County Ground Water. The MET-One particle count for the OCGW matrix sample was 389.3 particles/mL. A two-species optical fingerprint was generated using LW spiked with *Cryptosporidium* at 250 spheres/mL and OCGW. The Identification Rate was 23.3%, with a False Positive Rate of 0.027%, indicating better discrimination between *Cryptosporidium* and OCGW than *Cryptosporidium* and WPE compared to the results shown in Table 3.13. The Effective Patrol Volume Rate was 235.9 mL/hr, similar to the initial WPE measurements, but the problem with OCGW was the Interference Factor of 100%, indicating that *Cryptosporidium* could not be seen in this matrix with this MALS configuration. The particle count (for particles of size > 2 μm) for this sample was 389.3 particles/mL, compared to 70.1 particles/mL in the WPE sample above, suggesting that when particulate content is high, there can be a significant increase in the Interference Factor and therefore decrease in the ability to detect *Cryptosporidium*. Measurements with PSL spheres in OCGW were not performed. The failure of the ability to detect oocysts in this matrix prompted the decision to focus solely on a single variety of finished drinking water (WPE) as opposed to the other, even more challenging waters.

Table 3.15
Results of test of MALS instrument under same experimental conditions as Table 3.13, but different instrument configuration. *Cryptosporidium* oocysts are spiked at 228/mL in WPE. Results show that the previous result of high Interference Factor is repeated.

Target species/ matrix	PC/ *spike*[1]	Total # of detected events identified as:			ID Rate (%)	FP Rate (%)	EPVR (mL/hr)	EVM (mL)	Interf. Factor (%)
		WPE	*Crypto*	UNK					
Matrix[2] (WPE)	76.6	1333	5	11483		0.14	*1506.4*	*167.38*	
Crypto/LW[2]	*228/mL*[4]	79	417	674	35.6		46.2	5.13	0
Crypto/WPE[3]	*228/mL*[4]	1759	44	846					86.8

Notes:
1. Normal type indicates a particle count (PC) measurement of the sample on the MET-One particle counter, and italics indicates that no PC measurement was made and the number indicates the target spiking concentration.
2. Five 80 second measurements.
3. Four 80 second measurements.
4. Spiking concentration was lower than normal (here it is 228/mL) due to a temporary shortage of oocysts.

Test of Experimental Anomaly

One possible explanation of the high Interference Factor found in the initial *Cryptosporidium* spiking experiments in WPE (e.g., Table 3.13) was that it was an experimental anomaly.

Table 3.15 and Figure 3.9 show the results from measurements of *Cryptosporidium* spiked into WPE. This was essentially a repeat of the experiment shown in Table 3.13, but with a different instrument configuration. The MET-One particle count (for particles of size > 2 μm) for this WPE matrix sample was 76.6 particles/mL. An optical fingerprint was generated using LW spiked with *Cryptosporidium* at 228 oocysts/mL (lower than nominal due to limited availability of oocysts at the time) and WPE. The Identification Rate was 35.6%, with a False Positive Rate of 0.14%. The Effective Patrol Volume Rate was 46.2 mL/hr and the Interference Factor (IF) was 86.8%, essentially the same IF as achieved earlier (see Table 3.13). The left bar in Figure 3.9 is the mean number of *Cryptosporidium* events per time for the LW spiked samples, the center bar is the mean number of false positive events from the WPE sample, and the right bar is the mean number of *Cryptosporidium* events per time for the WPE spiked samples. Five 80 second measurements were used for the WPE and *Cryptosporidium* spiked into LW, while four 80 second measurements were used for the *Cryptosporidium*-spiked WPE. The error bars reflect one standard deviation around the mean.

In this experiment, a higher Identification Rate (IDR) was achieved, however, with an increased False Positive Rate (FPR). In addition, the EPVR was lower than previously found. However, the IF was nearly identical to the IF achieved earlier, so it was concluded that the low IF was likely not an experimental anomaly: it persisted through two experiments with two different instrument configurations made at two different times.

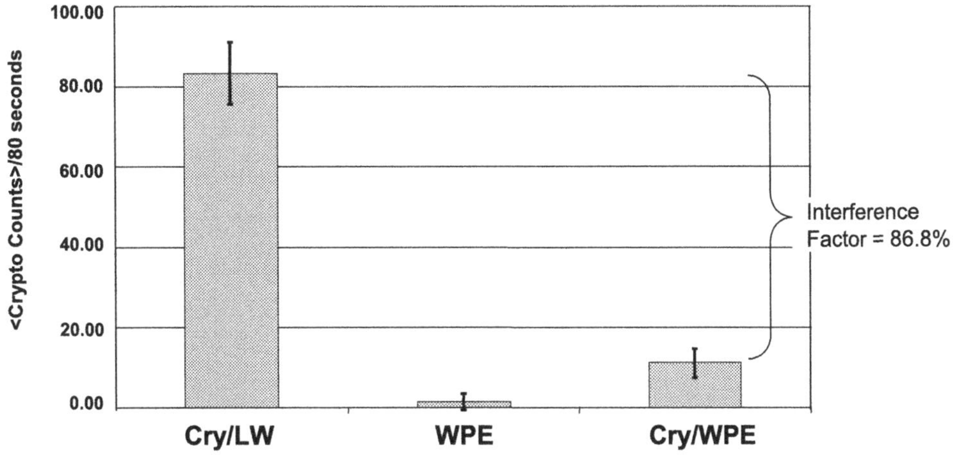

Figure 3.9 Results from measurements *Cryptosporidium* oocysts in WPE, based on data collected on the Gen-4 MALS system. The Interference Factor was 86.8% for *Cryptosporidium* spheres spiked at a concentration of 228 oocysts/mL.

Interference Factor as a Function of Spiking Concentration – for PSL Spheres

The next set of experiments investigated the effects of concentration on Interference Factor. Two sets of experiments were performed: first, a set with PSL spheres in order to eliminate the inherent variability associated with *Cryptosporidium* oocysts, and second, a set with *Cryptosporidium*. If the losses due to the WPE interference were due primarily to concentration, both PSL spheres and *Cryptosporidium* would have similar response as a function of concentration.

A three species optical MALS fingerprint was generated using a spiking concentration of 2 µm diameter PSL spheres at 500/mL, 4 µm PSL spheres at 500/mL and blank WPE. This optical fingerprint was then tested with both 2 µm PSL spheres and 4 µm PSL spheres spiked at concentrations of 100/mL, 500/mL and 1000/mL into purified water and WPE. The event rates for each spike were determined in purified water, then measured in drinking water. The ratio of each of these event rates – the event rate in drinking water divided by the event rate in purified water – the Interference Factor – is shown in Figures 3.10 and 3.11 In the 2 µm case the spike for the 500/mL case in purified water was suspect. There was no independent verification of spiking concentrations because the Met-One could not measure particles of this size reliably. In the 4 µm case, the spiking concentrations were measured using a MET-One particle counter.

In the 2 µm case, the Interference Factor was between 25% and 35% for the 1000/mL and 100/mL concentrations, but nearly 50% for the 500/mL concentration. The spike for the 500/mL concentration in purified water, which would have artificially raised the Interference Factor, was questionable. However, due to the particle size limitations of the MET-One particle counter, the spiking concentration was unverifiable.

In the 4 µm case, the Interference Factor is approximately 5% to 15%. The Interference Factor for 500/mL 4 µm is negative, but within experimental error, essentially 0%. Independent measurements of the 4 µm sphere spikes were performed using a MET-One particle counter. These results are typical – spiking concentrations for PSL spheres are generally within about 5% - 10% of the target concentration. The results of the particle counts are shown in Table 3.16.

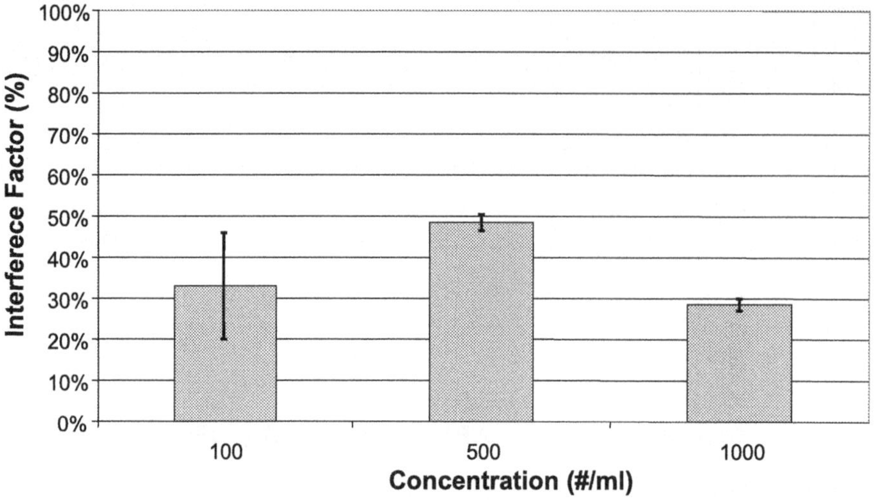

Figure 3.10 Results of test of Interference Factor as a function of spiking concentration for 2 μm diameter PSL spheres. "Interference Factor" was the decrease, in percent, of the number of 2 μm counts in WPE vs. purified water for the same spiking concentration. Error bars here reflect variations between six consecutive 2 minute data collections and reflect the variability within a given set of measurements.

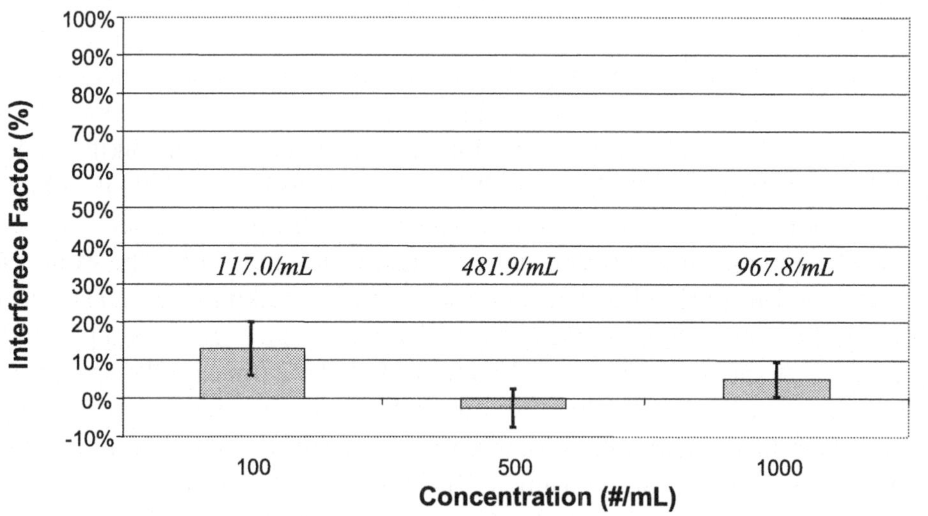

Figure 3.11 Results of test of Interference Factor as a function of spiking concentration for 4 μm diameter PSL spheres. The negative interference factor for the 500/mL case indicates that the rate of events/time measured in WPE exceeded the rate for 500/mL in purified water. Within error, however, the result is effectively zero. Error bars here reflect variations between six consecutive 2 minute data collections and reflect the variability within a given set of measurements. Spiking concentrations measured by the MET-One particle counter are shown in italics.

Table 3.16
MET-One particle counter measurements of 4 μm spiking concentrations, and particles/mL larger than 2 μm

	Nominal spike		
Particle/matrix	100 spheres/mL	500 spheres/mL	1000 spheres/mL
4 μm + LW	117.0	481.9	967.8
WPE (pre-spike)	23.8	22.3	23.3
4 μm + WPE	135.2	492.2	972.8

Measurements were performed on the spikes of 4 μm PSL spheres into purified (laboratory water – "LW") water, WPE pre-spike, and 4 μm PSL spheres into WPE post-spike, respectively. The target spiking concentrations were 100/mL, 500/mL, and 1000/mL.

These results indicated that within the spiking concentration range of 100 spheres/mL to 1,000 spheres/mL there was no apparent concentration effect on the Interference Factor. An effect would be manifested as a monotonic increase or decrease on the Interference Factor with concentration, most likely an increased as concentration increases due to self-interference. This was not supported by the data.

These results therefore suggest: (a) additional measurements using a wider variety of sphere sizes, at least down to 1 micron diameter were needed, and (b) there may be a size-dependent component to the high Interference Factors noted for *Cryptosporidium* in WPE. Since *Cryptosporidium* appears in light scattering curves to have intensities roughly between 1 μm and 2 μm PSL spheres, the increase in Interference Factor as the PSL spheres get smaller was relevant. If PSL spheres did not intrinsically agglomerate, then the event selection algorithms used in the analysis may have biased the results as the target particle size decreases. Correspondence with Duke Scientific, the vendor of the PSL spheres has indicated that the PSL spheres, which were designed to not self-agglomerate, may have in fact agglomerated to some unknown extent to other particles in the treated WPE water.

Interference Factor as a Function of Spiking Concentration – for *Cryptosporidium* Oocysts

A two-species optical MALS fingerprint was generated using 250/mL *Cryptosporidium* oocysts in purified water versus WPE.

Oocysts at concentrations of 50/mL, 100/mL, 150/mL, and 228/mL were spiked into purified water and WPE, respectively. The *Cryptosporidium* event rates for purified water were measured for each spiking concentration in purified water, and then measured for each concentration in drinking water. The ratio of these *Cryptosporidium* event rates – the Interference Factors – are shown in Figure 3.12.

These results indicated that the Interference Factors for *Cryptosporidium* in WPE were relatively constant as a function of concentration, similar to the results for PSL spheres. The Interference Factors for each spiking concentration in WPE were more than 80%, which was unsatisfactory and consistent with previous results for *Cryptosporidium.* While there was an apparent increase in the Interference Factor for the 100/mL and 150/mL spikes, this increase was not

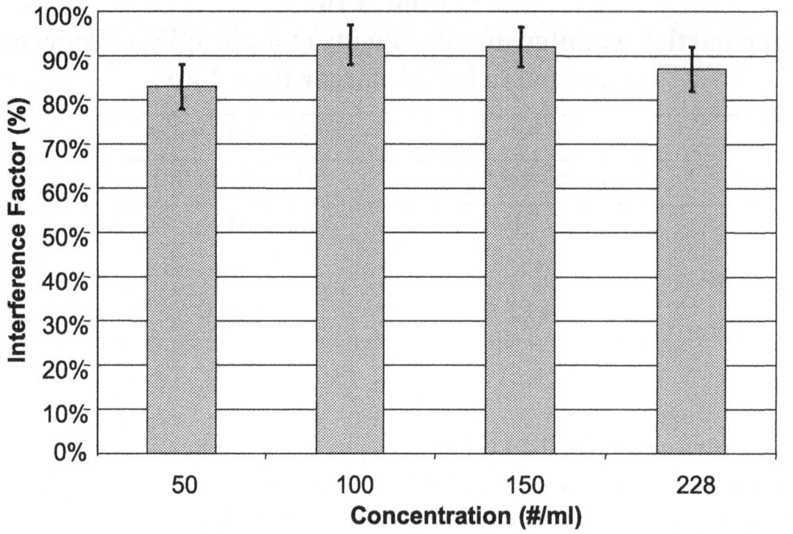

Figure 3.12 Interference Factors for *Cryptosporidium* in WPE as a function of spiking concentration. Error bars here reflect variations between individual two minute data collections and reflect the variability within a given set of measurements.

statistically significant, and there was no physical explanation for this increase. These measurements fell within one standard deviation of ±6%.

Measurement of the Effect of Water Chemistry on the Interference Factor for *Cryptosporidium*

In order to determine if the difference in the chemistry of the purified water and the drinking water was responsible for the high Interference Factors for *Cryptosporidium* oocysts in drinking water, an experiment was performed that used filtered WPE ("FWPE") as the matrix for generating the optical fingerprint. This experiment was designed to eliminate particulate interference in the fingerprinting matrix larger than about 0.1 µm without changing the basic chemical parameters of the water, e.g. pH, dissolved chemicals, etc. This test methodology is shown in Figure 3.13.

MALS optical fingerprints were generated using filtered WPE (FWPE) as the matrix rather than purified lab water. There were two steps to identification using the MALS technique: a "Training Phase" when hundreds or thousands of events are used to generate an optical fingerprint that distinguished between *Cryptosporidium* and the WPE (drinking water) matrix, and a Test Phase in which the WPE matrix was spiked with *Cryptosporidium*. If the water chemistry caused physical differences in the oocysts in purified water versus drinking water, and these differences caused the high Interference Factors, then using FWPE which nominally had the same chemistry, but different background particle content than WPE, would decrease the Interference Factor.

Filtered WPE (FWPE) was produced by passing WPE once through a 0.1 µm filter (Pall Gelman, supor-100). The FWPE was nearly as clean – based on the particle counter results – as purified laboratory water. Typical particle counts pre- and post-filtration, performed on a Met-One

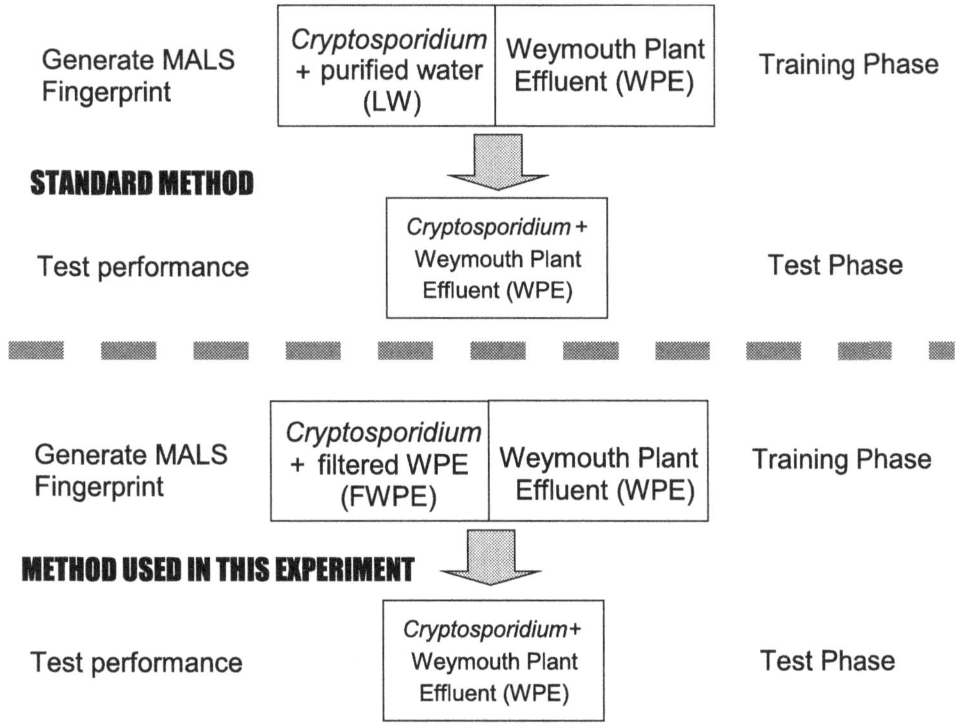

Figure 3.13 Modified optical fingerprint methodology used to test if chemistry affected the Interference Factor

Table 3.17
Typical particle count reductions generated by single-pass filtration of the WPE water through a 0.1 μm filter

	Matrix sample type	
WPE sample ID	WPE (particles/mL)	FWPE (particles/mL)
"WPE 071502 #1"	109.5	5.4
"WPE 071502 #2"	117.0	8.6
"WPE 071502 #3"	125.0	15.3
"WPE 071502 #4"	137.2	9.2

particle counter representing all particles of size greater than 2 μm, are shown in Table 3.17. For comparison, typical purified laboratory water (LW) generated counts from 0 to 5 particles/mL.

Three (3) liters of matrix samples (FWPE and WPE) were spiked with *Cryptosporidium* oocysts to a nominal concentration of 250 oocysts/liter and mixed for 30 minutes using a magnetic stir bar. The samples were then gravity fed through the MALS system at 100mL/min. One-half liter was flushed through the system prior to starting measurements in order to ensure that the mixture measured was the desired sample and not residual water left in the MALS system. At

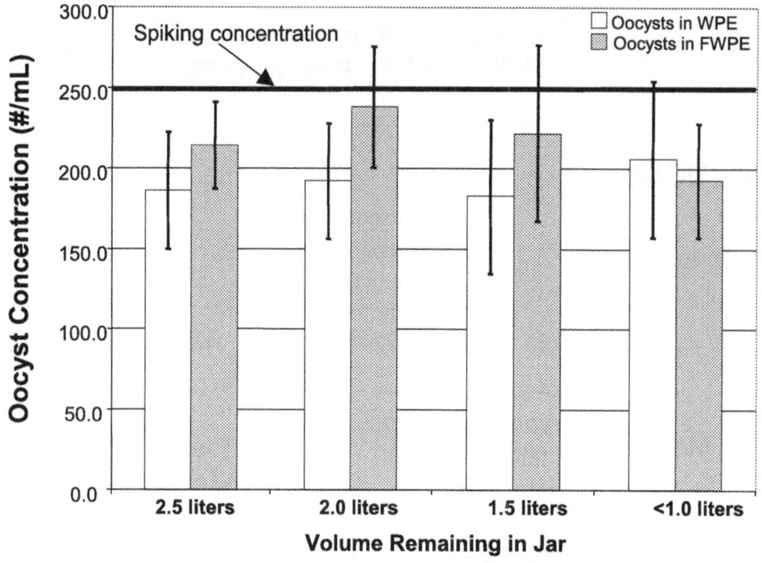

Figure 3.14 Results from the first set of 5/12 well IFA counts of *Cryptosporidium* oocysts spiked at 250/mL into FWPE and WPE at various points in the experiment. The first sample at 2.5 liters was immediately before the first MALS measurement, and the last measurement at "< 1.0 liters" was immediately after the last MALS measurement.

various times, corresponding to various remaining volume left in the aspiration jar, samples were taken out of the bottom of the system for immunofluorescence assay (IFA) analysis as a means to verify concentration. Limitations of obtaining enough *Cryptosporidium* oocysts created two problems with the performance of this experiment. First, it was not possible to generate a contemporaneous "control" using *Cryptosporidium* in purified lab water. Second, it was only possible to repeat this experiment twice.

The IFA analysis was performed as follows. 100 μL volumes were dried directly onto well slides, stained with an FITC conjugated antibody, and enumerated by microscopy. Microscopy was performed by three trained analysts with a random distribution of wells amongst the analysts. All wells were counted within 52 hours of receiving samples. Counting started using 12 wells in order to establish reasonable statistics, but after several measurements, this was switched to 5 wells for expediency. The variance in the measurements was unaffected by this change from 12 wells to 5 wells. The mean variance in the 12 well counts was 23.21% and was 23.6% for the 5 well counts. Results of the IFA counts are shown in Figures 3.14 and 3.15.

A two-species MALS optical fingerprint was generated using *Cryptosporidium* oocysts spiked to a concentration of 250/mL into filtered Weymouth Plant Effluent (FWPE) vs. unspiked Weymouth Plant Effluent (WPE). Results from two identical experiments of the event counts of *Cryptosporidium* events in FWPE as compared to the *Cryptosporidium* event counts in WPE are shown in Figures 3.16 and 3.17.

Despite the similarity in water chemistry between FWPE and WPE, the high Interference Factor persisted, similar to that shown in Figures 3.8 and 3.9 above. Therefore, water chemistry was not a primary cause of the high Interference Factors seen for *Cryptosporidium* in WPE.

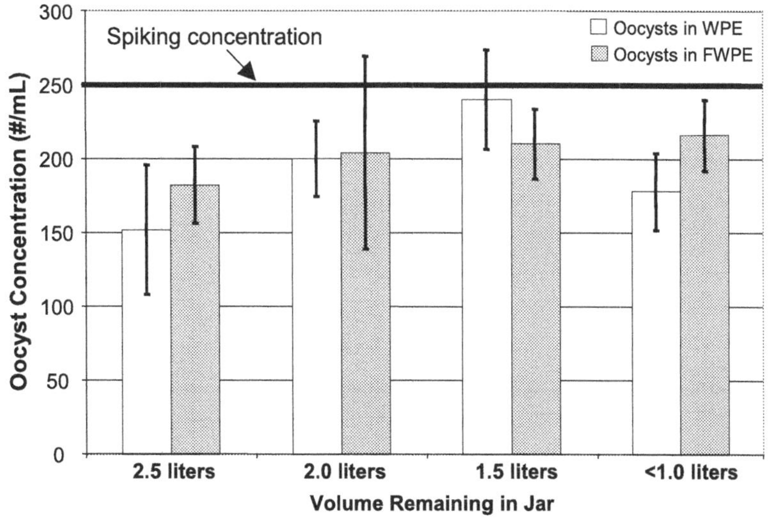

Figure 3.15 Results from the second set of 5/12 well IFA counts of *Cryptosporidium* oocysts spiked at 250/mL into FWPE and WPE at various points in the experiment. The first sample at 2.5 liters was immediately before the first MALS measurement, and the last measurement at "< 1.0 liters" was immediately after the last MALS measurement.

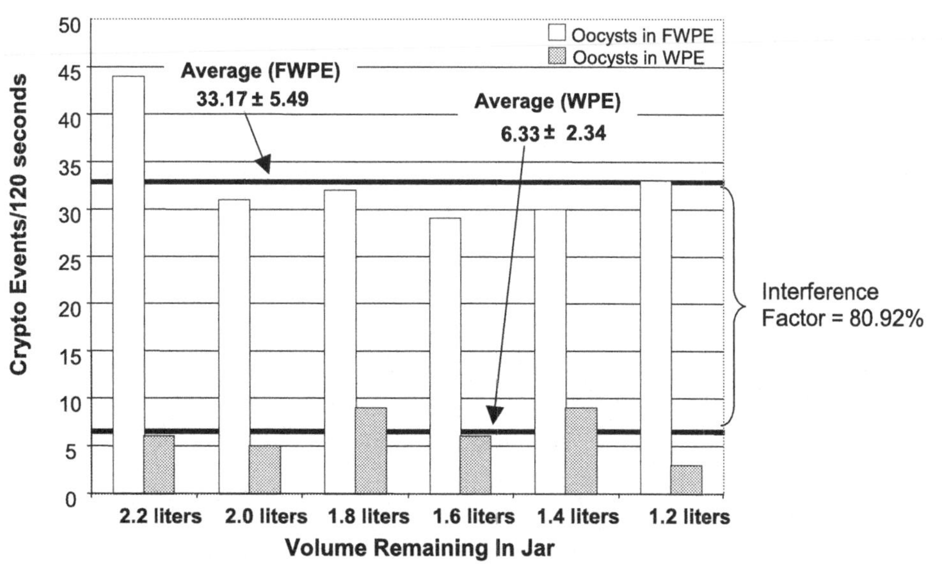

Figure 3.16 Results from the first set of MALS measurements of *Cryptosporidium* oocysts in FWPE vs. *Cryptosporidium* oocysts in WPE. FWPE and WPE were each spiked to a nominal concentration of 250/mL oocysts. The solid lines show the average of the six measurements. The Interference Factor was 80.92%, similar to the Interference Factor found with purified LW.

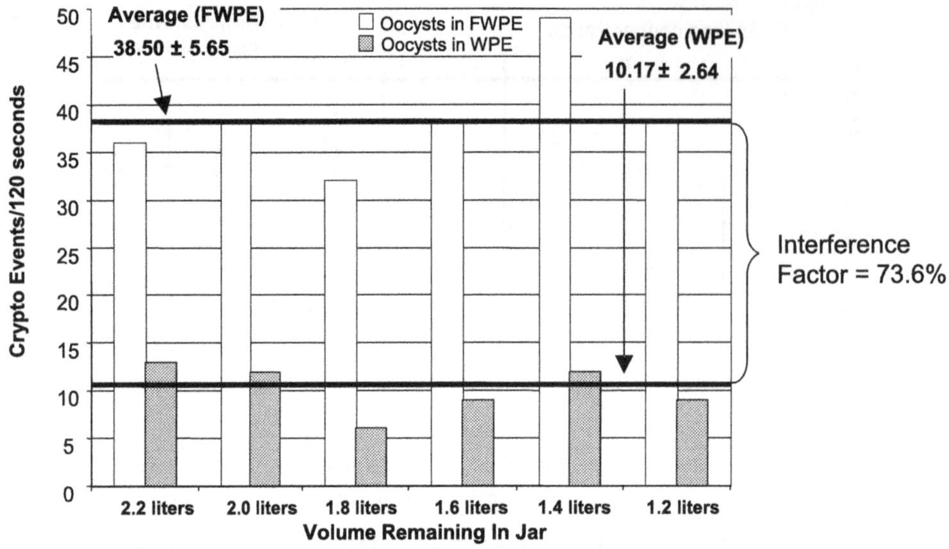

Figure 3.17 Results from the second set of MALS measurements of *Cryptosporidium* oocysts in FWPE vs. *Cryptosporidium* oocysts in WPE. FWPE and WPE were each spiked to a nominal concentration of 250/mL oocysts. The solid lines show the average and standard deviation of the six measurements. The Interference Factor was 73.6%, similar to the Interference Factor found with purified LW.

The IFA results indicated:

- The two repeated experiments produced generally consistent spiking counts.
- The spiking counts were low relative to the expected number to be recovered. Based on the flow cytometry counts provided by the Wisconsin State Laboratory of Hygiene, 250/mL should be spiked into the samples. The concentration as measured by IFA for Test #1 for all counts was 204.1 ± 19.4 per mL and for Test #2 was 197.8 ± 26.8 per mL. Losses could have occurred during the spiking process. The loss appeared to be about 20%. This may have been ascribable to spiking error or losses in the experimental process.
- The Interference Factors (one minus the ratio of oocysts counted in WPE to oocysts counted in FWPE) for the IFA were very low: 11.53% ± 10.24% in Test #1 and 5.17% ± 8.57% in Test #2. These were low enough, considering the high variability, to be considered almost negligible.
- There appeared to be no significant temporal trend in the concentration: so mixing was good, and there were no inhomogeneities in oocyst concentration seen in the results.
- Combining the previous two results, it can be concluded that – except for the approximately 20% presumed loss between theoretical spiking concentration and IFA measurements, which conceivably may have been due to spiking errors – experimental anomalies, e.g., sticking to tubes and test glassware, cannot account for the high Interference Factors of 85% to 90% previously observed.

The MALS results indicated:

- The two repeated experiments produced generally consistent results in terms of Interference Factors and oocyst counts per time. This was consistent with the IFA results.
- There appeared to be no significant temporal or volumetric trends of the results. This was also consistent with the IFA results.
- The Interference Factors were still relatively high, but slightly lower than previously measured using purified laboratory waters. The Interference Factor for Test #1 was 80.9% ± 7.73% and for Test # 2 was 73.59% ± 7.87%.
- From these measurements, it appeared that the chemical differences between purified water and WPE has a very slight effect if any on the Interference Factor, and could not (alone) explain the high Interference Factors.

Cryptosporidium Interference Factor in Various WPE Concentrations

The Interference Factor for purified laboratory water (LW) was defined as zero – the Interference Factor for treated drinking water (WPE) was measured in the range of 80% to 90%. In addition to investigating spiking concentration effects and water chemistry effects, Interference Factor as a function of WPE incremental blending into LW was measured. These results are shown here.

An experiment was performed to measure the Interference Factor as a function of WPE concentration as it was incrementally diluted with purified laboratory water. A two-species MALS optical fingerprint was generated using purified water spiked to 250 oocysts/mL versus WPE. A 5 liter purified laboratory water sample was spiked initially to 250 oocysts/mL. After the sample was measured and lowered to 2 liters, 300 mL of WPE and 700 mL of purified water were added to the sample, and an additional number of oocysts were added to bring the concentration back to 250/mL. This sample, now 10% by volume of WPE and 250/mL oocysts, was mixed for 6 minutes and tested. This process was repeated for higher concentrations of WPE of 20%, 30%, and 50%. Finally a sample of pure WPE was spiked at a concentration of 250 oocysts/liter. The results of this test are shown in Figure 3.18. The Interference Factor at 100% WPE concentration was 80.5%, slightly better than seen previously, but still high.

Figure 3.18 shows the linear relationship between the low WPE concentration results and the final data point at 100% WPE concentration. There may have been two or more effects occurring at different concentrations. These results were consistent with the phenomenon of oocysts sticking to matrix particles in that the Interference Factor was monotonically increasing with increasing concentration of WPE. However, further measurements and analysis were needed before this response was fully understood. It will be seen later that the Interference Factor was not due to oocysts sticking to matrix particles.

The results reported to this point eliminated four possibilities as potential causes of the high Interference Factor of *Cryptosporidium* observed in drinking water:

- First: Interference Factors were not an experimental anomaly. As the results above illustrate, several experiments were run after the initial measurement of high Interference Factors and in all cases, the Interference Factor in WPE was greater than about 80%.

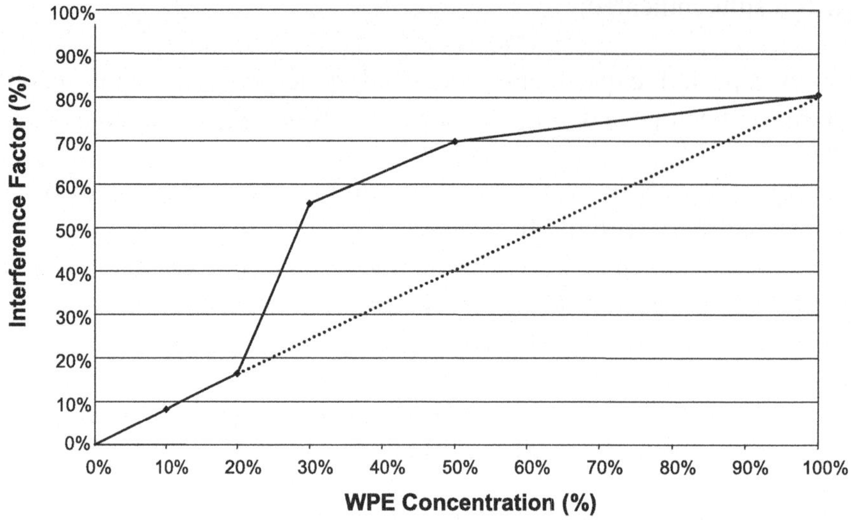

Figure 3.18 Interference Factor for *Cryptosporidium* oocysts as a function of WPE concentration. A line between the first three data points (0%, 10%, and 20% WPE concentration) and the last data point (100% WPE concentration) shows the linear relationship between the low concentration results and the very high concentration result.

- Second: Measurements of both PSL spheres and *Cryptosporidium* at various oocyst spiking concentrations revealed that concentration had no significant effect on Interference Factors.
- Third: Results suggest that the losses due to experimental procedures, such as mixing problems or sticking on input/output tubing were not significant. This was demonstrated by the high recovery rates (low Interference Factor) of using the IFA technique. At most there was a 10% to 20% loss due to experimental procedures. However, since this loss apparently occurred for all measurements, including spikes into purified laboratory water, this loss had no net effect on the Interference Factor.
- Fourth: The experiments using FWPE as the matrix with which to make *Cryptosporidium* fingerprints compared to purified laboratory water as the fingerprint matrix showed that there was little or no improvement in the Interference Factor. Therefore, water chemistry had no or minimal effect – at least for the limited treated water samples and spiking concentrations that were tested.

This left two possible remaining effects that could have caused high Interference Factors: analytical/electrical anomalies with the MALS system and/or agglomeration. One or both of these effects appeared to be occurring in the measurements.

Measurement of *Cryptosporidium* Interference Factors After System Modifications

In an effort to reduce the Interference Factor, two significant modifications to the MALS system were undertaken. There was a modification in the signal amplification scheme that reduced noise, and an improved feature extraction technique that was used to determine when an

Table 3.18
Measurement of the performance of the MALS instrument for *Cryptosporidium* oocysts at 250/mL in WPE after electronic and analytical modifications to the instrument

Target species/ matrix	PC/ *spike*[1]	Total # of detected events identified as: WPE	*Crypto*	UNK	ID Rate (%)	FP Rate (%)	EPVR (mL/hr)	EVM (mL)	Inter. Factor (%)
Matrix[2] (WPE)	126.0	742	17	531		1.32	*51.12*	*10.24*	
Crypto/LW[2]	*250/mL*	59	162	347	28.5		11.36	2.27	0
Crypto/WPE[2]	*250/mL*	795	93	843					42.6

Notes:
1. Normal type indicates a particle count (PC) measurement of the sample on the MET-One particle counter, and italics indicates that no PC measurement was made and the number indicates the target spiking concentration.
2. Six 120 second measurements.

electrical-optical response could be considered a target particle or noise. After these modifications were completed, an experiment was run to test the performance of the new MALS configuration.

Table 3.18 and Figure 3.19 show the results from measurements of *Cryptosporidium* spiked into WPE performed after electronic and analytical modifications were made on the MALS instrument. The Met-One particle count (for particles of size > 2 μm) for this WPE matrix sample was 126.0 particles/mL. An optical fingerprint was generated using LW spiked with *Cryptosporidium* at 250 spheres/mL and WPE. The Identification Rate was 28.5%, with a False Positive Rate of 1.32%. The Effective Patrol Volume Rate was 11.36 mL/hr, but the Interference Factor was 42.6%, a value significantly lower than the results reported earlier, and much closer to expectations. The left bar in Figure 3.19 is the mean number of *Cryptosporidium* events per time for the LW spiked samples, the center bar is the mean number of false positive events from the WPE sample, and the right bar is the mean number of *Cryptosporidium* events per time for the WPE spiked samples. Six samples were used for all cases. The error bars reflect one standard deviation around the mean.

Several experiments were run subsequent to the initial successful experiment to decrease the Interference Factor as shown in Figure 3.19. The results from these measurements are shown in Table 3.19. The dramatically improved Interference Factor is in the last column and shows the change after the MALS system was modified. Prior to the modifications, the Interference Factors were typically more than 80%, but after the modifications, the Interference Factors were about 10% to 50%, depending upon the event selection and optical fingerprint parameters. The Effective Patrol Volume Rate (EPVR) was markedly lower after the modifications, that is, the volume scanned by the system for *Cryptosporidium*, was significantly lower. The impact of the lower EPVR was that it would take longer to fully measure a fixed volume of water.

Two pairs of analyses of the same data using different analysis parameters are shown, illustrating how the analysis parameters were used to "tune" the performance of the system based on requirements. For example the "Post-Mod 3" second set of data showed an increase of Identification Rate and decrease in the Interference Factor at the expense of a higher False Positive rate, while the Effective Patrol Volume Rate stayed about the same.

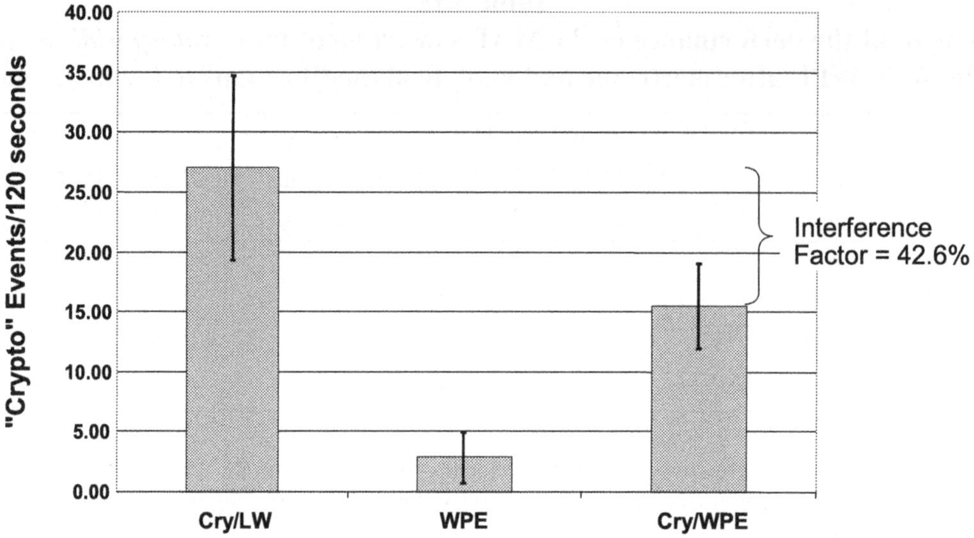

Figure 3.19 Interference Factor for *Cryptosporidium* oocysts in WPE, based on data collected on the MALS system after system improvements. The Interference Factor was 42.6% for *Cryptosporidium* spheres spiked at a concentration of 250 oocysts/mL.

Table 3.19
Summary of performance parameters for the detection of *Cryptosporidium* oocysts in WPE, before and after system modifications that dramatically improved performance

Experiment ID[1]	WPE sample ID	WPE PC[2] (#/mL)	IDR[3] (%)	FPR[4] (%)	EPVR[5]	IF[6] (%)
Pre-Mod 1	012502	70.1	1.4	0.044	213.48	86.9
Pre-Mod 2	022202-4	76.6	35.6	0.14	46.20	86.8
Post-Mod 1.1	093002-5	126.0	28.5	1.32	11.40	42.6
Post-Mod 1.2[7]	093002-5	126.0	11.2	0.31	12.48	37.1
Post-Mod 2	093002-8	109.1	12.4	0.31	7.56	50.9
Post-Mod 3.1	082602-E	102.0	34.5	0.77	2.88	20.7
Post-Mod 3.2[7]	082602-E	102.0	44.8	3.2	2.52	11.8

Notes:
1. "Pre-Mod" and Post-Mod" refer to the experiments taking place before and after, respectively, the modifications made to the system to improve performance, and the numbers correspond to the experimental set.
2. Particle count for the WPE sample, measured using the Met-One particle counter, particles/mL of size > 2 μm.
3. Identification Rate – the ratio of *Cryptosporidium* oocysts identified in LW to oocysts events detected in LW.
4. False Positive rate - the percentage of WPE matrix particles mis-identified as *Cryptosporidium*.
5. Effective Patrol Volume Rate – the effective volume patrolled per time (mL per hr) by the instrument.
6. Interference Factor – the percentage *Cryptosporidium* oocysts not detected in WPE matrix due to matrix interference.
7. Results from a second analysis on same data, using different event selection and optical fingerprint parameters.

Table 3.20
Example of stability of the Gen-4 MALS instrument for one month period

Date of measurement	Days from optical fingerprint generation	4 μm PSL sphere ID rate (%)
Jan. 2, 2002	-9	93.0
Jan. 7, 2002	-4	91.7
Jan. 11, 2002	0	95.6
Jan. 15, 2002	4	97.4
Jan. 16, 2002	5	95.0
Jan. 17, 2002	6	97.0
Feb. 1, 2002	21	97.7

SYSTEM STABILITY

The stability of the MALS system was tested by periodically spiking laboratory water with standard PSL spheres. The purpose of the stability measurements was to verify that the instrument was stable from day-to-day and to enable comparison of results performed on different dates under the same instrument and analysis configuration. Instrument configurations included changes to detector positions or swapping out of detectors that were not meeting specifications, electronics changes, and/or laser changes.

Stability was measured by generating an optical fingerprint using standards such as 1 μm, 2 μm, and 4 μm PSL spheres in LW. This optical fingerprint was then tested using 2 μm or 4 μm PSL spheres at various times. If the Identification Rate dropped off significantly, then the system required maintenance and the laser and detectors were inspected. Changes resulted in a new configuration that started the cycle over again. The results of this testing are shown below in Tables 3.20 to 3.21.

Table 3.20 shows the stability of the Gen-4 instrument soon after it was completed for a period of one month. Over the 31 day period shown, the system was stable, with 4 μm ID rates above 90%. The flow rate for this case were 10 mL/minute, prior to setting the standard flow rate to 100 mL/min.

Table 3.21 shows the stability of the Gen-4 instrument for more than two months at a later time. This figure shows the degradation of the system on 24 July 2002, which prompted maintenance to be performed on the system. The flow rate for this case is 100 mL/minute.

Figure 3.20 shows the stability of the MALS instrument for another 30 day period, in this case using an optical fingerprint generated with 2 μm PSL spheres and *Cryptosporidium* oocysts in LW. The ID Rate therefore was lower, around 65%. The MALS system was stable throughout this period with variance of 4.3%.

Table 3.21
Example of stability of the Gen-4 MALS instrument for more than two month period

Date of measurement	Days from optical fingerprint generation	4 µm PSL sphere ID rate (%)
May 16, 2002	0	96.7
May 27, 2002	11	95.7
May 28, 2002	12	94.3
Jun. 5, 2002	20	95.1
Jun. 17, 2002	33	95.1
Jun. 18, 2002	34	95.3
Jul. 10, 2002	55	93.2
Jul. 23, 2002	68	93.4
Jul. 24, 2002	69	79.5

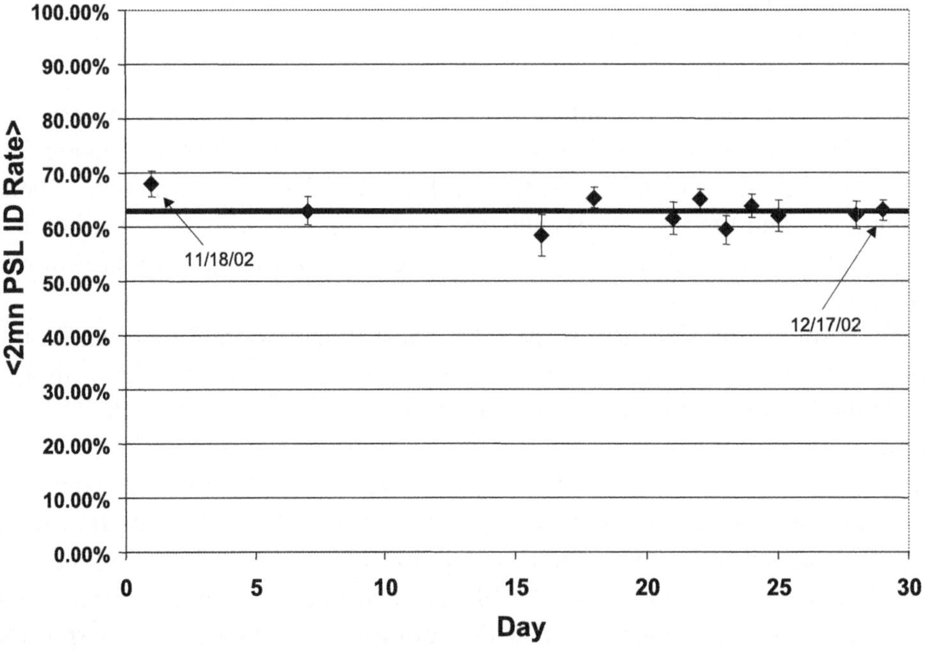

Figure 3.20 Stability of the MALS instrument for 30 day period as measured using 2 µm PSL spheres

ESTIMATED LIMIT OF DETECTION OF *CRYPTOSPORIDIUM* OOCYSTS IN LABORATORY WATER

The EPA has established a procedure for determining the level of detection application to Method 1623 – a grab sample technique. Since MALS was a continuous real-time technique, this methodology is not appropriate for MALS. Here a different approach was used for establishing the limit of detection (LOD) for the MALS technique, based on standard methodology used in the instrumentation industry. That is, the limit of detection was the lowest signal that was three standard deviations above the noise level (no oocysts were present) of the instrumentation.

For the purposes of this proof-of-concept effort, the "estimated" limit of detection (ELOD) for the MALS system in purified laboratory water (LW), based on the optical and electrical configuration used at the time of the measurements, was determined from a single serial dilution experiment and then verified with individual spiking experiments at a single high and several low concentrations.

Serial Dilution Experiment

A two species optical fingerprint was generated using 250/mL spikes in LW, with 2 μm PSL spheres and *Cryptosporidium* oocysts as the target particles. Then a serial dilution experiment using the approximate spiking concentrations of 250/mL, 125/mL, 63/mL, 32/mL, 16/mL, 8/mL, 4/mL, 2/mL and 1/mL was performed. Initially 250 oocysts/mL were spiked into three liters of LW. The water was mixed well using a magnetic stir bar for six minutes in order to improve homogeneity. 200 mL of the water was run through the system so that the initial water passing through the read-head was representative of the water in the aspiration jar and not residual LW from the tubing leading to the read-head. Five or six data files counting the number of *Cryptosporidium* oocysts detected per two minutes were collected, dropping the water level down to 1.5 liters. Laboratory water was then added to bring the volume back up to 3.0 liters and the process was repeated eight times, each time with an oocyst concentration of about half the previous concentration.

A control test was also run using the identical procedure as with the oocysts, but with LW only, in order to establish a LW baseline background. Despite attempts at generating particle-free water, it was impossible to eliminate completely all particles in the water. In general, some particles were always expected to be present, creating a low-level background noise level, even in LW.

Results are shown in Table 3.22. The first column shows the approximate concentration of the oocysts in the serial dilution, from the initial concentration of 250 oocysts/mL, diminishing in factors of two for each dilution, to 1 oocyst/mL. The next two columns show the mean *Cryptosporidium* counts in the two minutes files for the *Cryptosporidium* experiment and the LW baseline control, respectively. The fourth column shows the results from a statistical analysis performed on the two minute files, using the well-known t-test. In this case, the t-test tested the hypothesis that the means are the same for the *Cryptosporidium* sample and the LW baseline sample. If the probability was less than 1% (0.01), then the t-test failed and results from that concentration were deemed "detectable," as shown in the last column.

The results in Table 3.22 indicate that *Cryptosporidium* in laboratory water was detectable at least down to 8 oocysts/mL, but the LW background in this experiment became too noisy for the 4 oocyst/mL and 2 oocyst/mL concentrations for unambiguous detection. The *Cryptosporidium* counts were above the background, but not significantly so. At 1 oocyst/mL, the noise level in

Table 3.22
Results from a serial dilution experiment for the limits of detection of *Cryptosporidium* in laboratory water (LW)

Approximate concentration[1] (oocysts/mL)	Average *Cryptosporidium* counts[2] in 2 minutes		t-test probability[4]	Detectable[5]
	Cryptosporidium	LW baseline[3]		
250	57.4 ± 4.0	0.2 ± 0.5	2.4×10^{-6}	YES
125	27.4 ± 6.2	0.2 ± 0.5	2.9×10^{-6}	YES
63	19.2 ± 3.6	0.6 ± 0.9	8.9×10^{-5}	YES
32	10.0 ± 3.5	0.2 ± 0.5	0.0016	YES
16	4.4 ± 2.1	0.6 ± 0.6	0.0065	YES
8	3.4 ± 1.1	0.6 ± 0.6	0.0015	YES
4	2.6 ± 1.5	2.4 ± 1.1	0.41	NO
2	2.4 ± 1.1	1.4 ± 1.3	0.12	NO
1	2.6 ± 1.2	0.7 ± 0.8	0.0007	YES

Notes:
1. The initial concentration was 250 oocysts/mL, and the uncertainty in this concentration was about 10%. The uncertainty in each subsequent dilution increased due to compounded errors in volumetric measurements and spiking error.
2. The number of two-minute runs for each spiking concentration is either 5 or 6, except for the 1 oocyst/mL case, when water was run out of the aspiration jar. Ten two-minute runs were collected for this case.
3. A LW control was performed using identical procedures used for the *Cryptosporidium* serial dilution case, except no *Cryptosporidium* was added to the water initially.
4. The t-test probability that the mean *Cryptosporidium* counts of the *Cryptosporidium* spike and the LW control (no spike) are the same.
5. If the t-test probability was less than 1% (0.01), then the spike was deemed to be detectable in LW.

the LW was low enough for detection again, but the response had flattened out at this concentration. This type of response may indicate that noise from particles in the LW limited sensitivity.

The results below 8 oocysts/mL suggest that repeated experiments were desirable to derive a more statistically meaningful results, however, the multiple serial dilution technique was fundamentally limited because of the additive nature of the noise. Each subsequent dilution added noise and at the low concentrations, the artificial noise masked the signal from the oocysts. The advantage of the serial dilution technique was that the experiment could performed quickly. The disadvantage seen here was that the error became increasingly large as more and more dilutions were performed, so the sensitivity of the technique decreased naturally at lower concentrations. A better, but more time consuming method was to perform individual single dilutions – these results in LW are shown below.

The *Cryptosporidium* count flattened out below 4 oocyst/mL – this was consistent with noise due to the LW and the multiple dilutions creating interference. A graph showing the response of the Gen-4 MALS system between 250/mL and 8/mL is shown in Figure 3.21. A linear fit to the data is shown, extrapolated down to 0 oocysts/mL. Y axis error bars represent fluctuations in the measurements and are one standard deviation from the mean, x axis error bars are estimated error due to dilution and spiking error.

The serial dilution data exhibited a linear response as a function of concentration as expected.

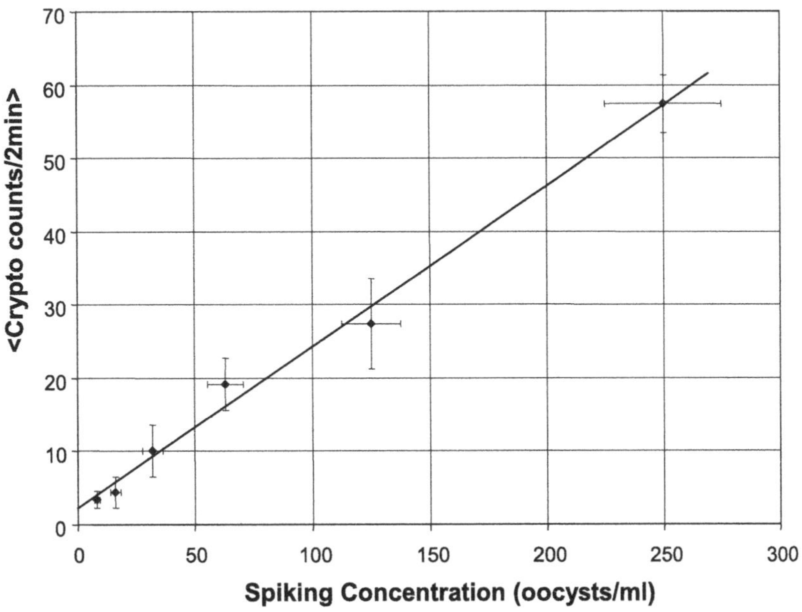

Figure 3.21 Average *Cryptosporidium* counts per two minute measurement as a function of spiking concentration based on a serial dilution test

Low Concentration Spiking Experiments

Several low concentration tests were performed in laboratory water in order to verify the serial dilution experiments and to explore extending measurements to lower concentrations of oocysts under more favorable noise conditions. Specific concentrations tested were: 250/mL, 10/mL, 1/mL, 0.1/mL, and 0.01/mL. Because the aspiration jars held only about 8.85 liters and there was some volume loss due to initial drainage through the read-head and some residual inaccessible water, a single test could last only about 70 minutes at 100 mL/min (for a total of seven liters tested) running through the read-head. The lowest concentration tested was 0.01 oocysts/liter. A companion set of measurements using 2 μm PSL spheres was performed using concentrations of: 250/mL, 10/mL, 1/mL, 0.1/mL.

Table 3.23 shows the single spike concentration experiments that were performed. A LW blank sample, using a full aspiration jar with no spike, was used for a background control.

Results from the single concentration experiments are shown in Table 3.24. In all cases, except for the 0.1/mL spike for 2 μm PSL spheres, the spiking concentration was detectable.

One important conclusion that was drawn from these results was that the estimated limit of detection is not simply a fixed value, but a function of time. This was due to the simple fact that the MALS system is an on-line continuous system that completely samples a fixed volume per time, which volume is dependent upon the water flow rate and the feature extraction parameters.

The t-test probability shown in Table 3.24 was calculated using the two-minute counts for each species. In the case of the 250/mL spike it was eight two-minute files, and for the other spiking concentrations, 32 to 38 two-minute counts were used to form the distribution.

All of the *Cryptosporidium* spikes from 250/mL to 0.01/mL shown in Table 3.24 were found to be "detectable" based on t-test performed on spiked samples vs. blank with 2 minute

Table 3.23
Single spike concentration experiments performed to determine the limit of detection of 2 μm PSL spheres and *Cryptosporidium* in purified lab water

Species	Concentration	Number of 2-minute files	Total run time (min)
2 μm PSL	250/mL	8	16
2 μm PSL	10/mL	34	68
2 μm PSL	1/mL	34	68
2 μm PSL	0.1/mL	32	64
Cryptosporidium	250/mL	8	16
Cryptosporidium	10/mL	36	72
Cryptosporidium	1/mL	35	70
Cryptosporidium	0.1/mL	34	68
Cryptosporidium	0.01/mL	35	70
LW Blank	0/mL	38	76

counts. However, the 0.1/mL result for *Cryptosporidium* was suspect because there was no decrease in mean count from the 1/mL spike. This could be due to either spiking error or heterogeneity in the sample. Also, the 0.01/mL result for *Cryptosporidium* was suspect because less than three total events are expected. This could have been due to spiking error, LW background noise, or a heterogeneous sample.

The 2 μm PSL sphere spikes from 250/mL to 1/mL shown in Table 3.24 were determined to be "detectable" also based on a t-test statistic. The exception was the 0.1/mL experiment. In this case, a simulation was run using the same data concatenated twice more, for a total of 192 minutes. Under these conditions, the t-test probability was 0.003, sufficient for detection. Based on this result, for concentrations less than 1/mL for 2 μm PSL spheres, measurement times of more than 70 minutes were necessary to make a detection based on t-test statistics.

Minimum Time of Detection as a Function of Concentration

Using the data that generated the results shown in Table 3.24, another technique, Poisson change point analysis (c.f., Fitzgerald and Rooney, 1996), was used to determine the minimum time of detection for *Cryptosporidium* oocysts in purified laboratory water. The time series of the events measured for a given concentration was concatenated onto the time series of events for the blank (no spike) simulating a step function change in concentration. See Figure 3.22. The Poisson change point analysis (PCPA) provided an optimal method to determine: (a) when a transition from blank to spiked sample occurred, and (b) an estimated mean value for the number of events per time in the blank and the spiked sample.

Results of applying PCPA to the *Cryptosporidium* spikes of concentration 250/mL, 10/mL, 1/mL, 0.1/mL and 0.01/mL are shown in Figures 3.23 through 3.27.

Table 3.24
Results from a series of spiking concentrations for the detection of *Cryptosporidium* and 2 μm PSL spheres in laboratory water (LW)

[Spike][1] #/mL	*Cryptosporidium* spikes				2 μm PSL sphere spikes			
	2 μm[2]	*Crypto*[3]	t-test prob.	Detect[4]	2 μm[5]	*Crypto*[6]	t-test prob.	Detect[4]
250	0.25 ± 0.5	71.9 ± 8.8	2.9×10^{-8}	YES	178.9 ±1 5.2	4.0 ± 1.9	2.9×10^{-9}	YES
10	0.0 ± 0.0	2.4 ± 1.5	1.1×10^{-10}	YES	9.62 ± 3.73	1.03 ± 1.03	1.2×10^{-16}	YES
1	0.029 ± 0.17	1.47 ± 1.41	1.6×10^{-7}	YES	1.32 ± 1.34	0.35 ± 0.54	1.0×10^{-6}	YES
0.1	0.21 ± 0.5	1.53 ± 1.13	3.2×10^{-9}	YES	0.06 ± 0.25	0.16 ± 0.37	0.08	NO
0.01	0.0 ± 0.0	0.59 ± 0.74	6.5×10^{-5}	YES				NA
0[7] (LW Blank)	0.0 ± 0.0	0.03 ± 0.16[8]	NA	NA	0.0 ± 0.0	0.03 ± 0.16[8]	NA	NA

Notes:
Shaded area indicates that the measurement was not made.
1. Spiking concentration for 2 μm PSL spheres or *Cryptosporidium* oocysts, in number per mL.
2. False positive rate, in mean number of detections per two-minute measurement of 2 μm PSL spheres in a spike of *Cryptosporidium* oocysts in LW.
3. Mean number of identifications of *Cryptosporidium* oocysts in a spike of *Cryptosporidium* in LW in two minutes.
4. Determination of ability to detect the target species (2 μm PSL spheres or *Cryptosporidium* oocysts) based on the t-test probability of less than 1% (0.01).
5. Mean number of identifications of 2 μm PSL spheres in a spike of 2 μm PSL spheres in LW in two minutes.
6. False positive rate, in mean number of detections per two-minute measurement of *Cryptosporidium* oocysts in a spike of 2 μm PSL spheres in LW.
7. Measurement of baseline LW blank sample – equivalent to spiking concentration of 0 particles/mL.
8. One event identified as *Cryptosporidium* occurred in this 76 minute test of purified laboratory water.
"NA" – Not applicable

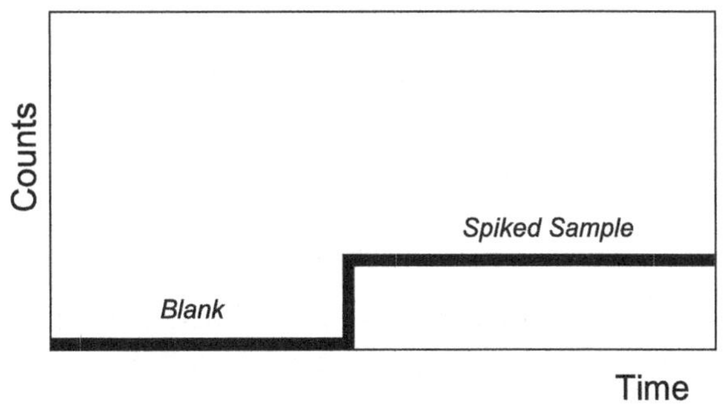

Figure 3.22 A step function change in concentration is simulated and Poisson change point analysis is used to determine the minimum time of detection and the estimated mean event rate for a spike in laboratory water.

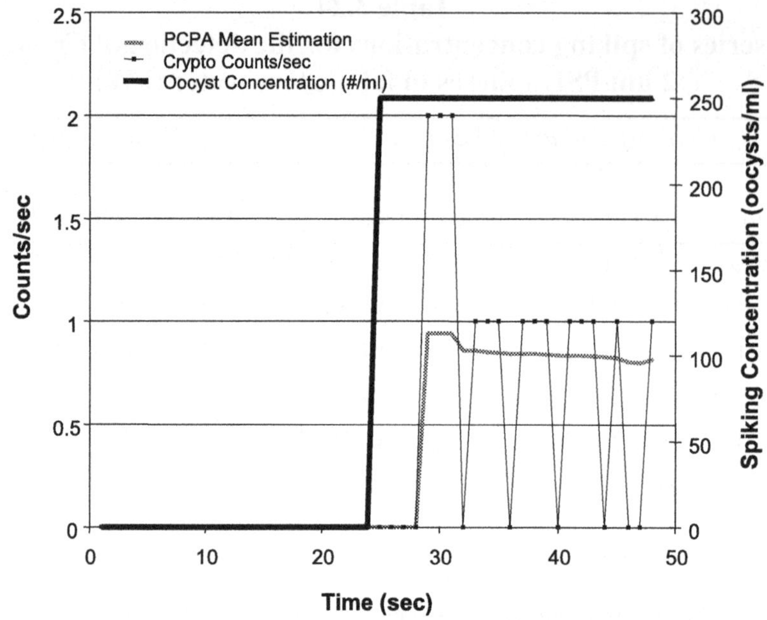

Figure 3.23 Poisson Change Point Analysis (PCPA) of spike of 250 oocysts/mL in LW. Time is in seconds. The spike occurred at 25 seconds, and was detected five seconds later by PCPA.

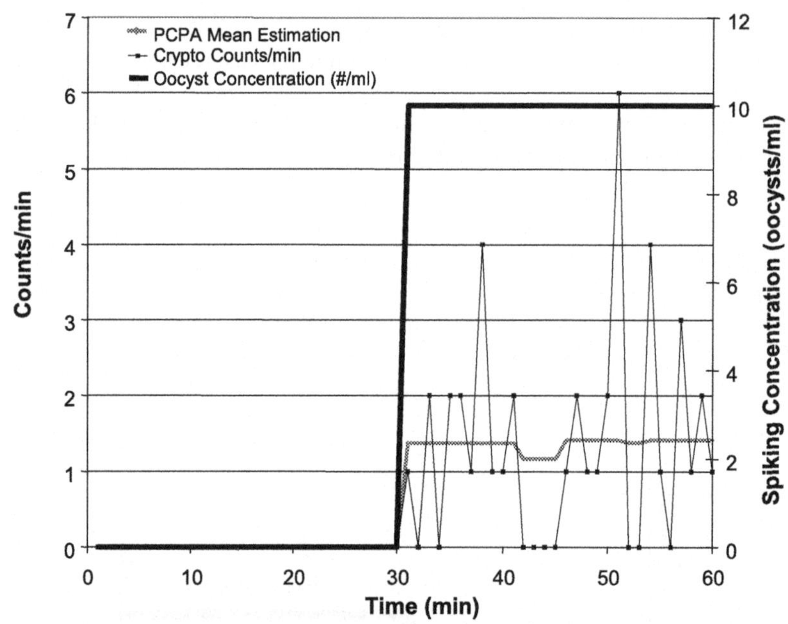

Figure 3.24 PCPA of spike of 10 oocysts/mL in LW. Time is in minutes. The spike occurred at 30 minutes, and was detected within one minute by PCPA.

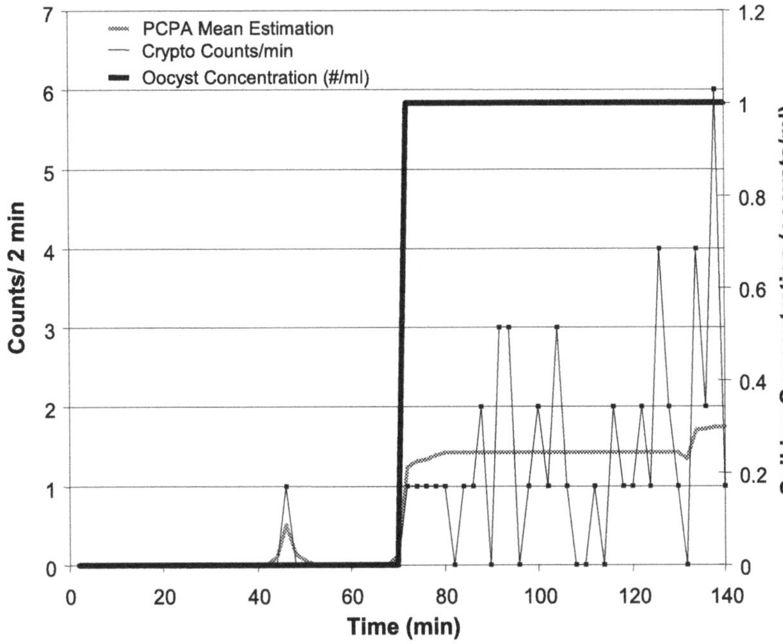

Figure 3.25 PCPA of spike of one oocyst/mL in LW. Time is in minutes. The spike occurred at 70 minutes, and was detected within two minutes by PCPA.

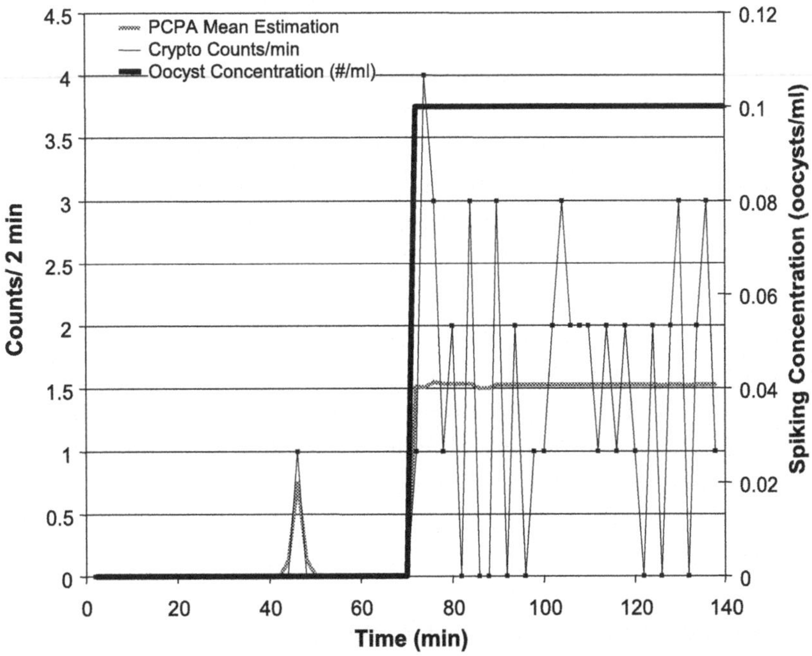

Figure 3.26 PCPA of spike of 0.1 oocyst/mL in LW. Time is in minutes. The spike occurred at 70 minutes, and was detected within two minutes by PCPA.

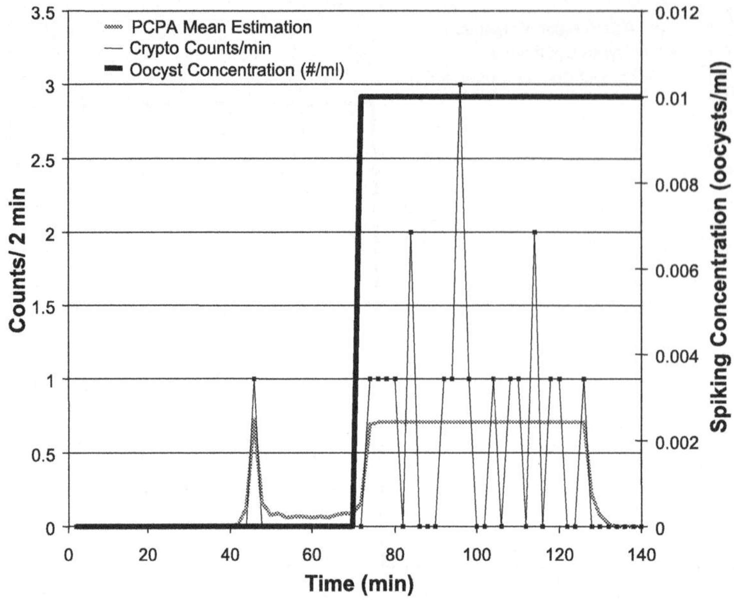

Figure 3.27 PCPA of spike of 0.01 oocyst/mL in LW. Time is in minutes. The spike occurred at 70 minutes, and was detected within two minutes by PCPA.

A summary of the results of the PCPA is shown in Table 3.25. Concentration is shown in oocysts/mL, detection time is in minutes, and mean counts correspond to the estimated mean provided by the PCPA.

For the 250 oocysts/mL concentration, the data were broken into one second segments and the number of events per second were used as input to the PCPA. The detection time was determined to be five seconds for this particular measurement, but in this particular case no events were detected for four seconds and the average number of events per second for these data was 0.7 events/second. It is likely that the detection time was somewhat lower than five seconds (0.12 minutes), perhaps as low as one or two seconds.

For the 10 oocysts/mL concentration, the data were broken into one minute segments, and the detection time was determined to be one minute. It is likely that the detection time was somewhat less than one minute.

For the remainder of the concentrations, the data were maintained in the normal two-minute files. Estimated detection times are shown as less than two minutes for the 1 oocyst/mL and 0.1 oocyst/mL case, and less than four minutes for the 0.01 oocyst/mL case. However, it was noted above that the 0.1 oocyst/mL and 0.01 oocyst/mL spiking cases were likely unreliable and required repeated measurements.

Calculated and Theoretical Best Limit of Detection

The results reported for the serial dilution test (Table 3.20) and the low concentration spikes (Table 3.23) suggest that the limit of detection for the MALS instrument was not limited by false positive rates in purified laboratory water (LW), but rather by the amount of water sampled per time. Based on this, the estimated limit of detection (ELOD) as a function of time was calculated from the performance parameters achieved in LW. The Effective Patrol Volume Rate

Table 3.25
Summary of results for the PCPA for various spiking concentrations of oocysts in LW

Spike (oocysts/mL)	Detection time (min)	Mean counts (2 min)
250	~ 0.1	101.1 ± 5.9
10	< 1	2.8 ± 0.2
1	< 2	1.42 ± 0.05
0.1	< 2	1.53 ± 0.05
0.01	< 4	0.71 ± 0.03

(EPVR) depended upon the existing configuration of the instrument. The EPVR for the analysis shown here was 14.4 mL/hr. The Identification Rate was a measure of the efficiency in the instrument correctly identifying oocysts as they generated events in the read-head. The Identification Rate for the analysis shown here was 70.0%. The Interference Factor (IF) was the percentage reduction in properly identified *Cryptosporidium* events in a real waster sample relative to a purified water sample. It was defined therefore as 0% in laboratory water.

The formula for estimated limit of detection (ELOD) is:

$$ELOD(t) = (t * EPVR * IDR * (1\text{-}IF))^{-1} \qquad (3.3)$$

where t is time in minutes, PVR is the patrol volume rate in mL/min, IDR is the Identification Rate, DY is the Detection Yield, and IF is the Interference Factor which is defined as 0% in LW. ELOD is the minimum concentration of oocysts in #/mL detectable for a measurement taken over time t.

Figure 3.28 shows the result of this calculated and estimated limit of detection. In addition, Figure 3.28 shows the results of the PCPA analysis from Table 3.25. Note that the single spike results were generally consistent with the calculated ELOD. Included is a fit to the PCPA results for the results through 1 oocyst/mL. The fit showed a steeper slope than the calculated ELOD, and suggested better performance than predicted at the lower concentrations. This supposition requires further study and additional experiments.

Also shown for illustration in Figure 3.28 is the maximum theoretical performance for the MALS instrument using a single read-head. The maximum theoretical performance was calculated using the following assumptions:

1. The Identification Rate and Detection Yield were both 100%.
2. The water speed through the system was increased by a factor of five.
3. The laser geometry was modified to increase the Patrol Volume by a factor of two.

Using these assumptions, the maximum theoretical performance for a single MALS read-head was about 100 times better than what was achieved in LW.

In Figure 3.28, the thin solid line is the calculated estimated limit of detection for the Gen-4 MALS configuration used in this experiment. The circles represent the results of the Poisson

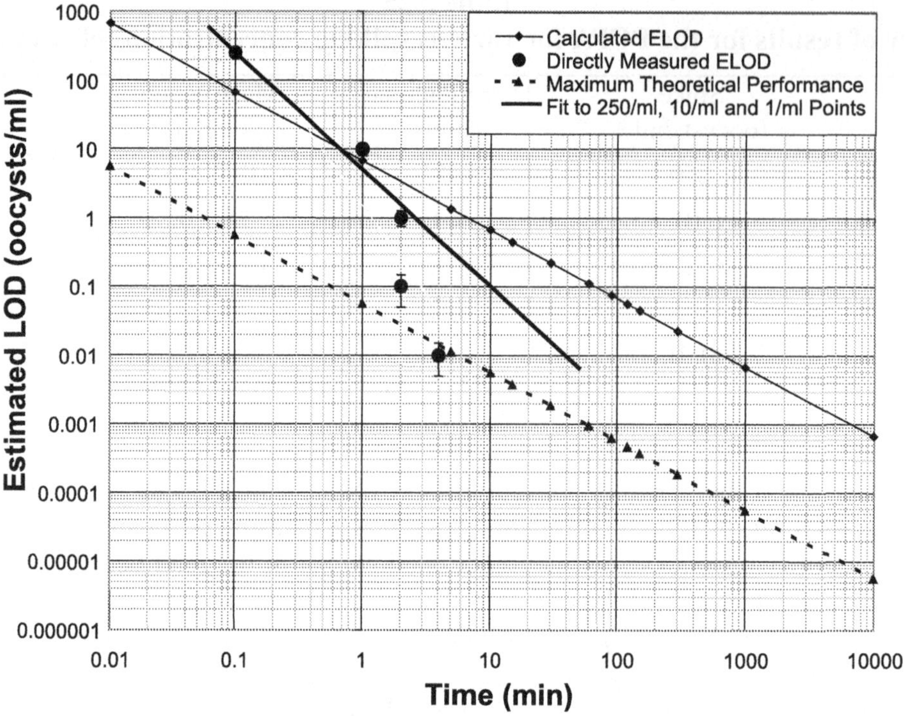

Figure 3.28 Estimated limit of detection for Gen-4 MALS in for *Cryptosporidium* in purified lab water

change point analysis (PCPA) which determined the time required to detect a specific concentration. The solid line is a fit to the three largest concentrations. The dotted line represents the best possible performance for a single Gen-4 read-head.

ESTIMATED LIMIT OF DETECTION OF *CRYPTOSPORIDIUM* OOCYSTS IN FINISHED DRINKING WATER

The "estimated" limit of detection (ELOD) for the MALS system for finished drinking water (WPE), based on the optical and electrical configuration used at the time of the measurements, was determined from a single serial dilution experiment. Unlike the purified laboratory water case, single spiking experiments were not performed due to time limitations. In order to minimize the variability of the matrix on the results, a single 10 liter matrix water sample was used. While a single experiment was insufficient to provide a statistically meaningful ELOD, this experiment was performed to provide an indication of the order of magnitude of the limit of detection for *Cryptosporidium* oocysts in drinking water, in this case Weymouth Plant Effluent.

A two species optical fingerprint was generated using a 250/mL spike of *Cryptosporidium* in LW and WPE. Then a serial dilution experiment using the approximate spiking concentrations of 250/mL, 83/mL, 27/mL, 9/mL was performed. A blank control of WPE with no spike was also analyzed.

Initially 250 oocysts/mL were spiked into three liters of WPE. The water was mixed well using a magnetic stir bar for six minutes in order to improve homogeneity. 200 mL of the water

Table 3.26
Summary of results for the serial dilution of *Cryptosporidium* in WPE

Samples[1]	Approximate concentration[2] (oocysts/mL)	Mean identifications of *Cryptosporidium* and WPE in spiked sample: *Cryptosporidium* counts[3]	WPE [3] counts	t-test probability[4]	Detectable[5]
7	250	6.7 ± 2.8	26.3 ± 4.6	0.00054	YES
7	83	6.1 ± 3.3	28.7 ± 7.0	0.0021	YES
7	28	3.9 ± 1.5	29.1 ± 5.4	0.00038	YES
10	9	3.6 ± 1.0	28.9 ± 6.1	8.2×10^{-7}	YES
6	0 (WPE Blank)	0.7 ± 0.5	28.4 ± 12.2	NA	NA

1. Each sample is a 2-minute measurement.
2. Concentration is approximate because serial dilutions introduce significant uncertainty in the actual concentration. Concentration is estimated assuming homogeneous mixing and accurate dilutions.
3. Mean and standard deviations of the counts for the number of samples measured.
4. Test of the hypothesis that the measured results were from the same distribution as the zero spiking results. The t-test result measures the probability that the two distributions are the same.
5. A threshold of less than 0.01 (greater than 99% probability) indicates that the estimated concentration of *Cryptosporidium* oocysts is detectable in drinking water.

were run through the system so that the initial water passing through the read-head was representative of the water in the aspiration jar and not residual from the tubing leading to the read-head. Seven files counting the number of *Cryptosporidium* oocysts detected per two minutes were collected, dropping the water level down to 1.0 liter. WPE was then added to bring the volume back up to 3.0 liters and the process was repeated three times, each time with an oocyst concentration of about one-third the previous concentration.

The results from this test are shown in Table 3.26. The first column shows the number of two minute samples measured for each concentration. The second column shows approximate concentration of the oocysts in the serial dilution, from the initial concentration of 250 oocysts/mL, diminishing in factors of three for each dilution, to 9 oocysts/mL. The next two columns show the mean *Cryptosporidium* counts and the mean WPE counts in the two minute files, respectively. The fifth column shows the results from a statistical analysis performed on the two minute files, using the well-known t-test. In this case, the t-test tested the hypothesis that the means are the same for the *Cryptosporidium* spike in WPE sample and the WPE (unspiked) baseline sample. The test failed and the means were determined to be different and thus the results from that concentration were deemed "detectable" if the probability was less than 1% (0.01). In all cases, down to 9 oocysts/mL, *Cryptosporidium* was detectable in WPE.

The results shown in Table 3.26 are plotted in Figure 3.29. Also shown in the solid black line is a linear regression fit to the data, including the WPE blank control. These results, like the LW results, indicate that the limit of detection for the MALS instrument was not limited by false positive rates in drinking water, but rather by the amount of water sampled per time. Thus the estimated limit of detection (ELOD) can be calculated from the performance parameters achieved in WPE. The Effective Patrol Volume Rate (EPVR) depended upon the existing configuration of the

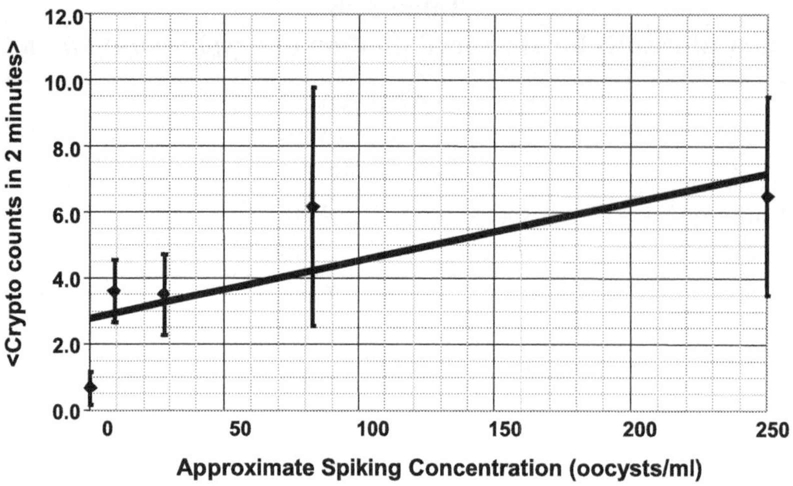

Figure 3.29 Response of the Gen-4 MALS system measured in average *Cryptosporidium* counts per two minute measurement as a function of spiking concentration for WPE

instrument. The EPVR for the analysis shown here was 3.6 mL/hr. The Identification Rate for the analysis shown here was 34.5% and the Interference Factor was 20.7%. Based on these numbers, and using Equation 3.1 to calculate the ELOD, Figure 3.30 shows the ELOD for this sample of drinking water as a function of time and also shows the maximum theoretical performance for a single read-head, assuming Identification Rate was 100%, the Effective Patrol Volume Rate was 240 mL/hr, the Interference Factor is 0%, and the water flow rate increased by a factor of five.

DISCRIMINATION OF PHYSICALLY ALTERED OR TREATED OOCYSTS

A series of experiments were performed to evaluate the ability for MALS to discriminate between viable *Cryptosporidium* oocysts and treated or physically altered oocysts. Most current identification methods, e.g. immunofluorescent assay (IFA), could not directly distinguish between infectious vs. non-infectious oocysts because IFA was dependent upon protein identification and not specifically sensitive to morphological changes that may occur due to treatment. The MALS technique has been shown to be extremely sensitive to morphological differences, both external and internal in microparticles. For example, changes in diameter of merely 5 nm in a 1 μm diameter polystyrene latex sphere was detected by the MALS technique. (c.f. Wyatt, 1975)

Five *Cryptosporidium* oocyst sample types were tested:

1. Untreated viable oocysts
2. Heat inactivated oocysts
3. Ozone inactivated oocysts
4. Excysted oocysts
5. UV disinfected oocysts

The methods used for generating these oocysts are discussed in Chapter 2, Materials and Methods. Samples were shipped and measured in two sets, two days apart. This was done so that fresh samples would be measured by MALS immediately after treatment thus minimizing any

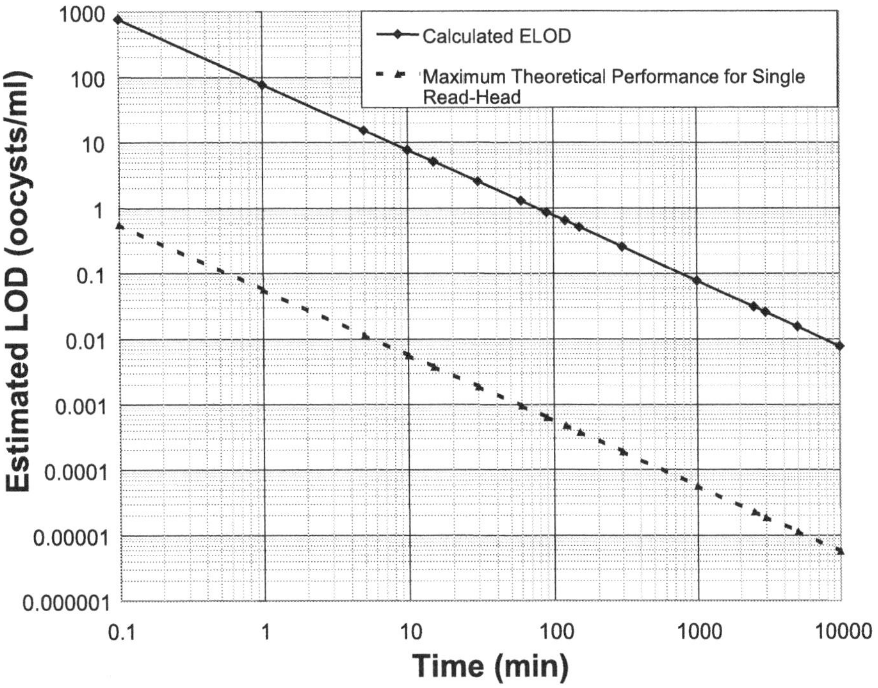

Figure 3.30 Estimated limit of detection for MALS for *Cryptosporidium* in a sample of drinking water (WPE)

time-related effects. Set 1, measured on the first day, included heat inactivated, ozone inactivated, and untreated viable oocysts. Set 2, measured on the second day, included excysted oocysts, UV disinfected oocysts, and untreated oocysts. All of the samples were measured within 12 hours of treatment.

Samples were tested for their infectivity to cell culture (see Materials and Methods) to ensure that the treatment techniques were effective. These results are shown in Table 3.27.

All MALS measurements were made in purified laboratory water. Except for Set 1 untreated control *Cryptosporidium* oocyst sample "C," which was used to generate the optical fingerprint, the measurements were done blind, that is, it was not known which samples were treated or untreated when tested.

A MALS optical fingerprint was constructed using 2 µm PSL spheres in LW and sample "C" untreated *Cryptosporidium* oocysts in LW, each spiked at 250/mL in LW. This optical fingerprint was used in the remainder of the analyses. The performance parameters for this optical fingerprint are shown in Table 3.28. A total of 1549 events were collected, the *Cryptosporidium* Identification Rate was 32.8% and the False Positive Rate (2 µm PSL spheres incorrectly identified as *Cryptosporidium*) was 0.53%.

Two methods were used to determine if the MALS technique could discriminate between the species. The first method was qualitative – observation of the raw light scattering patterns of each of the species. The second method involved generating pair-wise optical fingerprints of the treated/excysted samples and the untreated *Cryptosporidium* measured on the same day and comparing the Wilk's lambda from these pairs of data. Wilk's lambda is a measure of discrimination of distributions using multiple analysis of variance (MANOVA) (c.f., Cooley and Lohnes, 1971).

Table 3.27
Measured infectivity of oocysts under various treatments

Sample identification	Total number of oocysts in sample	Treatment	Log inactivation
A	2.16×10^6	80°C, 45 minutes	≥3.72[1]
B	2.72×10^6	Ozone (CT=160 mg.min/L)	≥3.72[1]
C	2.4×10^6	Control	NA[2]
D	1.17×10^6	Excystation (83.3%)	0[3]
E	3.12×10^6	Medium pressure UV (44.57 mJ/cm^2)	3.5[4]
F	3.04×10^6	Control	NA[2]

Notes:
1. Expected amount of kill (at least)
2. ID_{50} of control oocysts = 128
3. Not expecting to see any kill here – in fact, these would have had higher infectivity
4. Lower kill than expected

Table 3.28
Performance of optical fingerprint used for test of discrimination of treated vs. untreated oocysts

Total events	*Cryptosporidium* IDR	2 µm PSL IDR	Total *Crypto* events	Total 2 µm events	False Positive Rate	False Positive Events
1549	32.84%	66.06%	609	940	0.53%	5

Figures 3.31 through 3.36 show the raw data for the six samples which were measured. Each graph shows 100,000 scans, or eight seconds of continuous data along the x-axis, and the electrical response of the Gen-4 MALS system on the y-axis. All of the figures are on the same scale.

Visual inspection of the MALS raw data curves from the six oocyst samples revealed:

1. The day 1 untreated control oocysts and the day 2 untreated control oocysts looked very similar, as expected.
2. The ozone inactivated oocysts appeared to have slightly higher scattering amplitudes than the controls.
3. The excysted sample showed markedly higher scatter amplitudes than the controls.
4. There was no visible difference between the heat-treated, UV-inactivated and control oocysts.

Based on visual inspection alone, it was possible to tell the excysted sample from the controls, perhaps possible to distinguish the ozone treated sample from the controls, but the other samples appeared indistinguishable from the controls.

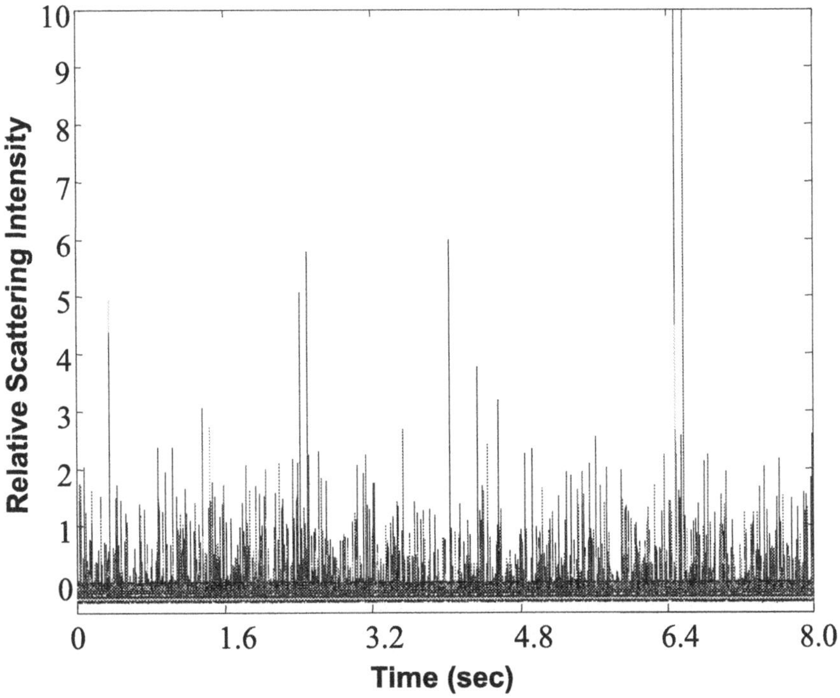

Figure 3.31 Raw MALS data from heat inactivated oocysts, Sample A

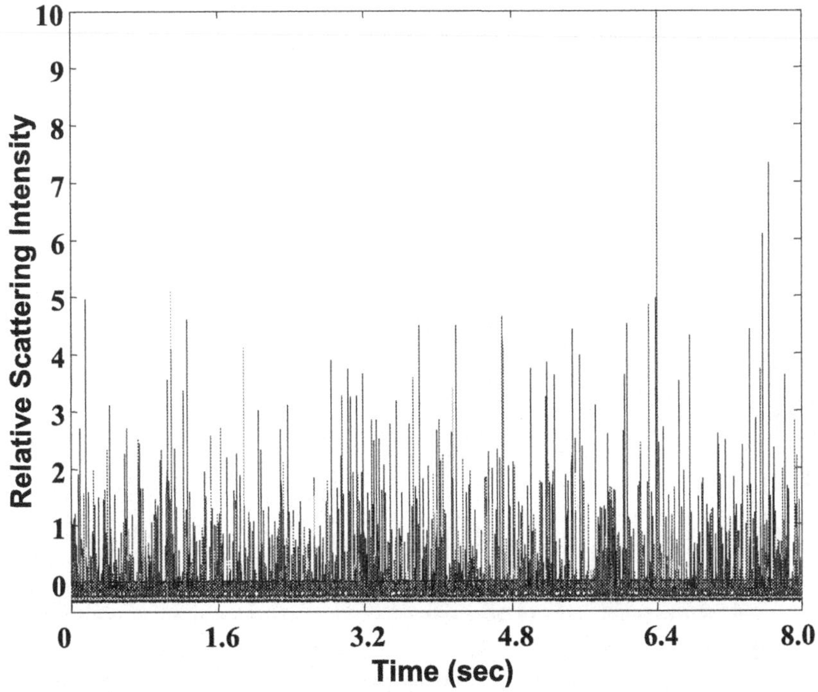

Figure 3.32 Raw MALS data from ozone inactivated oocysts, Sample B

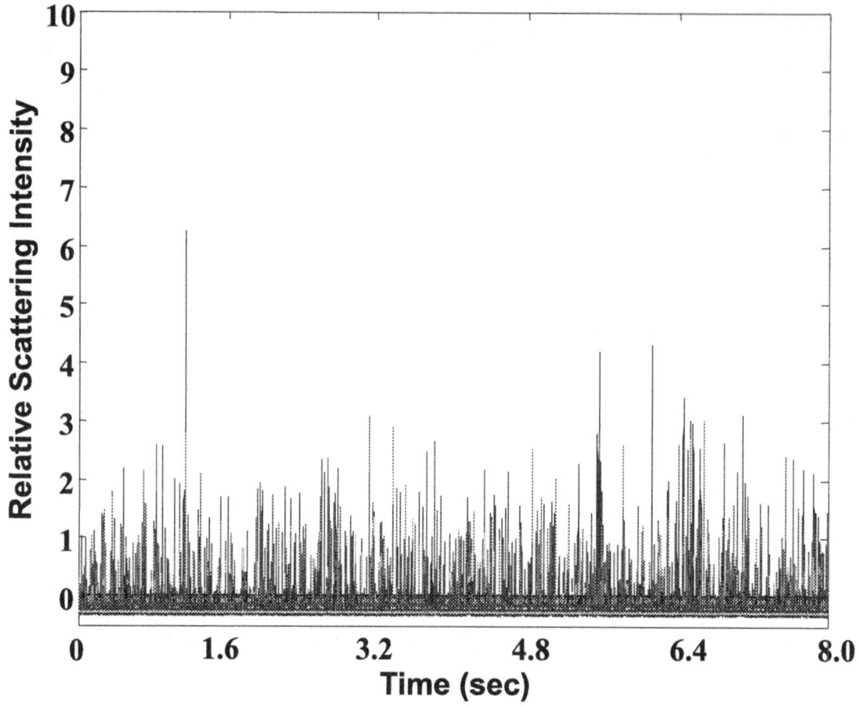

Figure 3.33 Raw MALS data from untreated oocysts, Sample C

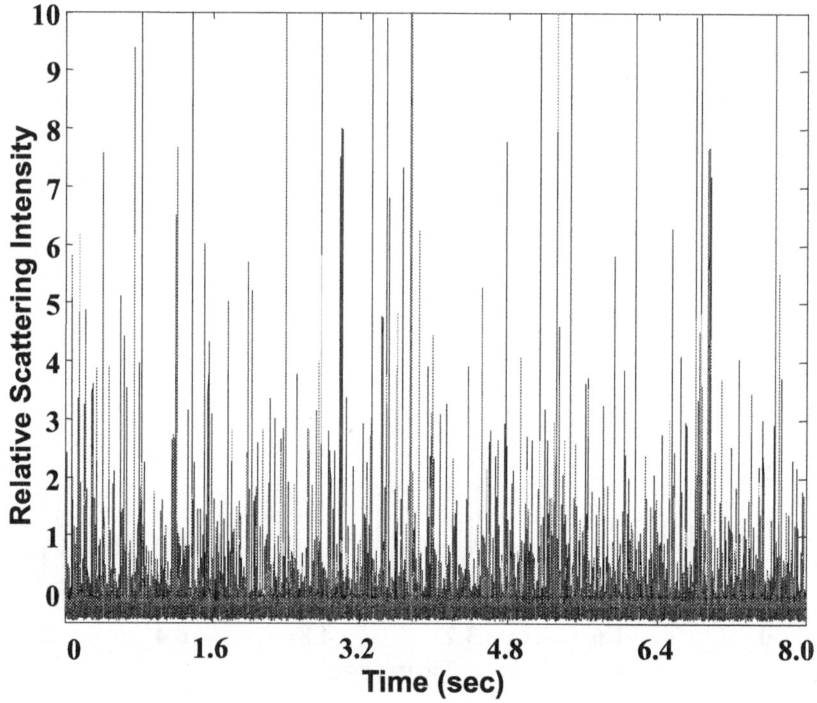

Figure 3.34 Raw MALS data from excysted oocysts, Sample D

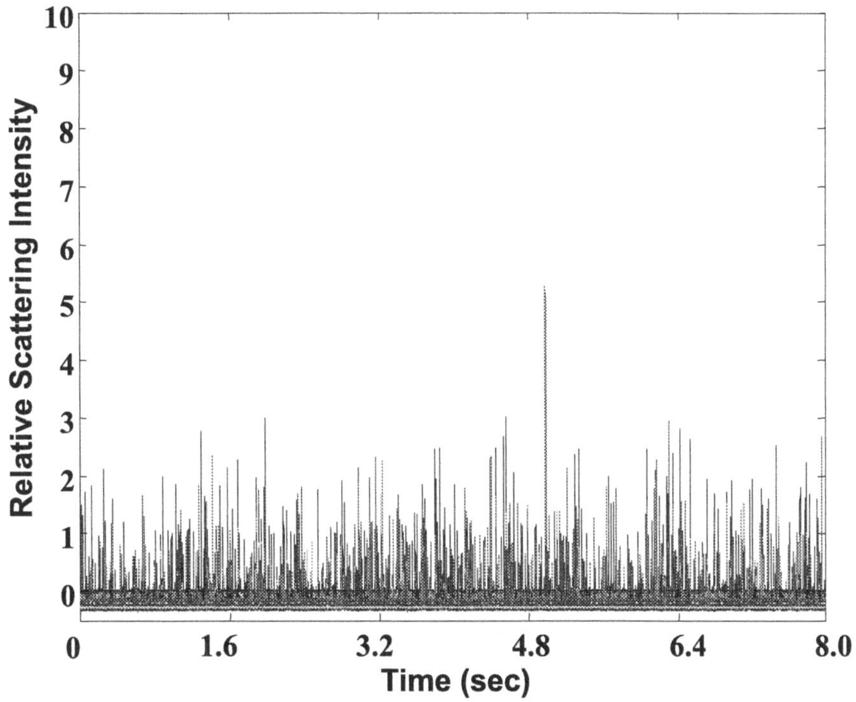

Figure 3.35 Raw MALS data from UV inactivated oocysts, Sample E

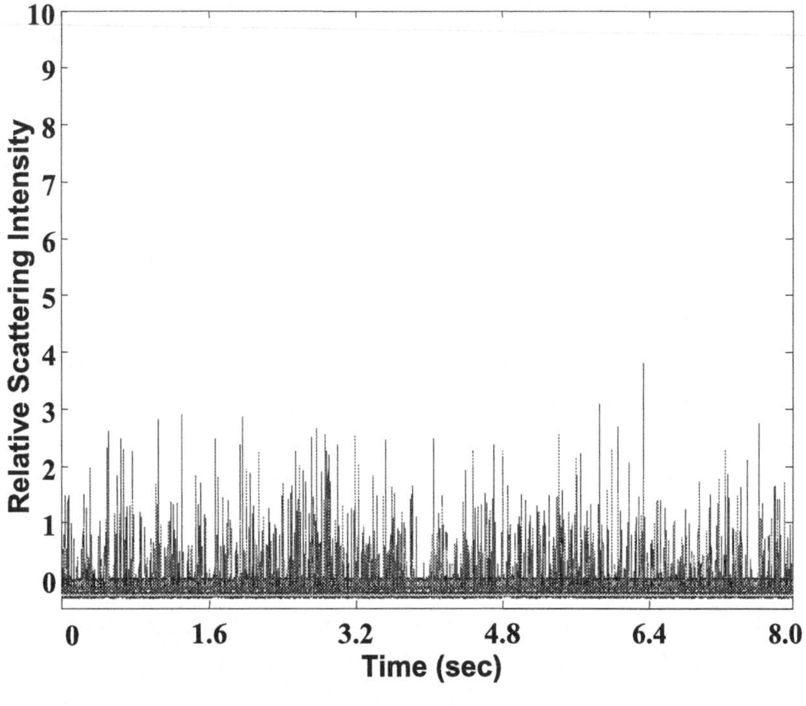

Figure 3.36 Raw MALS data from untreated oocysts, Sample F

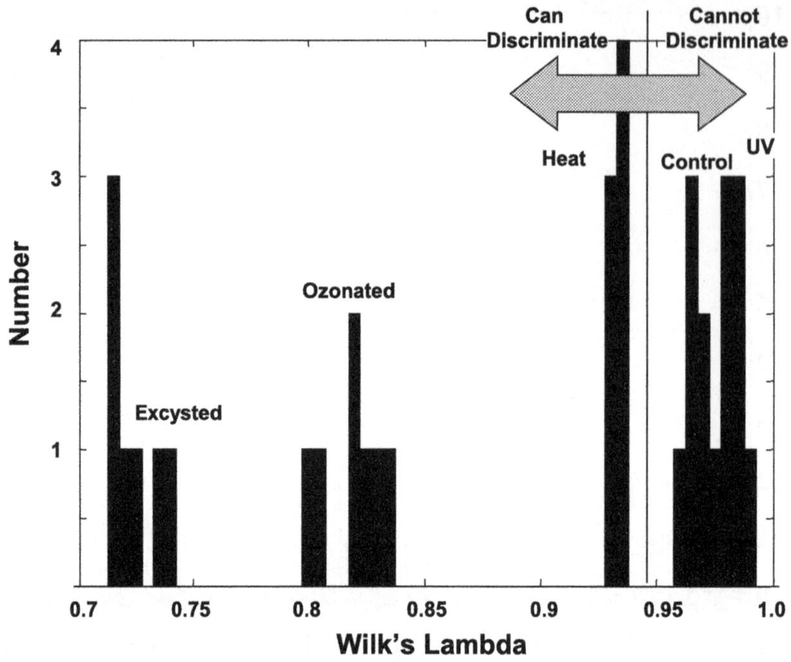

Figure 3.37 Distribution of Wilk's lambda for pairs of oocyst samples. "Control" corresponds to Day 1 viable oocysts vs. Day 2 viable oocysts. Otherwise, the indicated sample was compared against the viable oocysts from the same day.

The second method used to discriminate between the oocyst samples was by use of the Wilk's lambda generated by MANOVA. Eight two-minute measurements were made of each oocyst sample in purified laboratory water. MANOVA was performed pair-wise on treated/excysted oocysts versus the controls measured on the same day. A sensitivity test was performed to determine how many two minute measurements were necessary to get a stable Wilk's lambda – the result was that four two minute measurements sufficed. A distribution of Wilk's lambda was then generated by comparing randomly selected sets of four measurements from each treated/excysted oocyst sample against four randomly selected control measurements. Seven Wilk's lambdas per sample pair were generated and the distribution of the results is shown in Figure 3.37. The results indicate that visually, all of the treated/excysted samples except the UV sample were distinct from the control, which had a higher mean Wilk's lambda and overlapped the control distribution.

The distributions were also compared by means of the t-test to determine if they were from the same distribution (the null hypothesis) or not. It was assumed for this analysis that the distributions were normal. Detectability was defined by two criteria. First, the t-test results must be less than 1% (0.01), indicating that there was less than a 1% probability that the Wilk's lambda distribution was from the same distribution. Second, the mean of the treated sample distribution must be less than the control, which was generated by comparing day one untreated oocysts versus day two oocysts. The results of this t-test detectability criterion is shown in Table 3.29.

For the results shown in Table 3.29, the use of day one untreated oocysts versus day two untreated oocysts as the control distribution was conservative, since other pair-wise comparisons were done with treated samples from the same day. The Wilk's lambda for intra-day control-control comparison was greater than 0.99, while the inter-day comparison of the controls was 0.967. An

Table 3.29
Results from the measurement of oocysts indicating the ability of MALS to discriminate between viable oocysts and oocysts treated by various treatment techniques

Sample pair-wise comparison	Mean Wilk's lambda	t-test	Able to discriminate?
Day 1 Untreated vs. Day 2 Untreated (control)	0.9662 ± 0.0047	NA	NA
Day 1 Untreated vs. Day 1 Heat Treated	0.9329 ± 0.0014	1.8×10^{-7}	YES
Day 1 Untreated vs. Day 1 Ozonated	0.8196 ± 0.0125	2.0×10^{-9}	YES
Day 2 Untreated vs. Day 2 Excysted	0.7239 ± 0.0108	4.2×10^{-12}	YES
Day 2 Untreated vs. Day 2 UV	0.9838 ± 0.0028	8.2×10^{-12}	NO
Day 1 Untreated vs. Day 1 Untreated	0.99317	NA	NA
Day 2 Untreated vs. Day 2 Untreated	0.99241	NA	NA
Day 1 Untreated vs. 2 μm PSL Spheres	0.33091	NA	YES

example of untreated control oocysts versus 2 μm PSL spheres is shown for comparison, and the mean Wilk's lambda was 0.331.

The results shown in Table 3.29 were consistent with the visual observation that the excysted sample was significantly different than the other samples. The excysted sample versus the untreated control had the lowest Wilk's lambda, indicating the highest degree of discrimination of all of the pairs of treated samples, 0.7239.

In Figure 3.37, it can be seen that the control and the UV inactivated sample distributions were close and overlapped. However, the t-test performed on the pair generated a very low probability that these distributions came from the same distribution (i.e. the null hypothesis is rejected), but since the UV sample has a larger Wilk's lambda, it was determined to be not detectable.

MATRIX WATER ANALYSIS

Standard Water Quality Measurements

All of the water matrices used for this project were tested for water chemistry and particle counts using standard methods discussed in Chapter 2. The water chemistry tests were performed on all samples initially, but were performed only occasionally afterward. The water chemistry (at least to the extent for finished waters), e.g. TDS, TSS, TOC, played a minimal part in the performance of MALS in drinking water matrices for the detection of *Cryptosporidium* oocysts. Particle counts proved to be a more important independent measure of water quality for MALS, and particle counts were performed in conjunction with each MALS measurement. Particle counts (size greater than 2 μm) are shown with each matrix experiment in the results.

Table 3.30 shows the results from the standard water quality measurements on the various matrix waters used in this project.

The WPE samples in which the predominant number of measurements were performed provided a relatively stable water chemistry matrix over the period shown in Table 3.30. The turbidity was between 0.05 and 0.07 ntu; the TDS was between about 500 and 600 mg/liter; the TSS was generally below instrument detection limits, except for a single sample that had a TSS of 0.1 mg/liter;

Table 3.30
Water quality measurements from top set (Set 1), the first set of samples collected in 2001 for MALS testing; bottom set (Set 2), samples collected from Metropolitan (Weymouth Plant Effluent) in 2002

Sample[1]	Collection date	Turbidity (ntu)	TDS (mg/L)	TSS (mg/L)	TOC (mg/L)	Particle count (#/mL)[2]
PST Lab[3]	7/26/01	0.05	11	ND	0.1	2[4]
WPE	7/25/01	0.07	541	ND	2.81	108.8 ± 25.3
CRW	7/25/01	1.61	585	1.4	3.27	7709.2 ± 2403.6
IRGW	8/17/01	0.29	243	ND	0.74	615.2 ± 120.7
OCGW	8/17/01	0.13	1080	ND	1.76	532.0 ± 545.0
SPW	8/27/01	1.7	260	2.1	3.49	6898.4 ± 2045.5
WPE	4/18/02	0.07	510	ND	2.82	17.8 ± 0.8
WPE	4/9/02	0.07	503	ND	3.2	23.2 ± 1.2
WPE	2/22/02	0.05	560	0.1	2.77	80.8 ± 3.6
WPE	2/5/02	0.07	DNM	DNM	DNM	65.0 ± 6.6

Notes:
1. The sample abbreviations have the following meanings:
   - "PST Lab": the low particulate laboratory water ("LW") generated by PointSource Technologies, used for generating the *Cryptosporidium* optical fingerprint.
   - WPE: Weymouth filtration plant effluent.
   - CRW: Colorado River Water, collected at the Metropolitan Diemer filtration plant.
   - IRGW: Irvine Ranch Ground Water, collected at the corner of Redhill Ave. & Barranca Parkway, Santa Ana.
   - OCGW: Orange County Ground Water, collected at Prado wetlands, well "CBWM 0600664."
   - SPW: California State Project Water, collected at Metropolitan's Live Oak reservoir in La Verne, CA.
2. Mean ± one standard deviation, based on five repeated measurements.
3. The measurements for PST Lab water likely reflect instrument sensitivity limits.
4. A single measurement was performed for the PST Lab water on this date.

"ND" – Not detectable
"DNM" – Did not measure.

and the TOC was in the range of 2.7 to 3.2 mg/liter. In contrast to the water chemistry parameters, the particle counts for WPE samples had a relatively large variance. The lowest counts are shown in Table 3.30, around 17.8 particles/mL (> 2 μm), and ranged occasionally up to 150 particles/mL. Because MALS depends upon the ability to measure one particle at a time as it passes through the laser beam, the particle content of the matrix water directly affected the performance of MALS in terms of Identification Rate and Recovery Rate, which will be seen in the results that follow.

MALS Measurements

In addition to performing standard water quality measurements on the matrix water samples, the samples were consistently tested and characterized using the MALS instrument. Figures 3.38 through 3.43 show various water types as measured in the MALS system. These data were raw data, that is, the MALS system electrical response as a function of time as particles in the various matrices

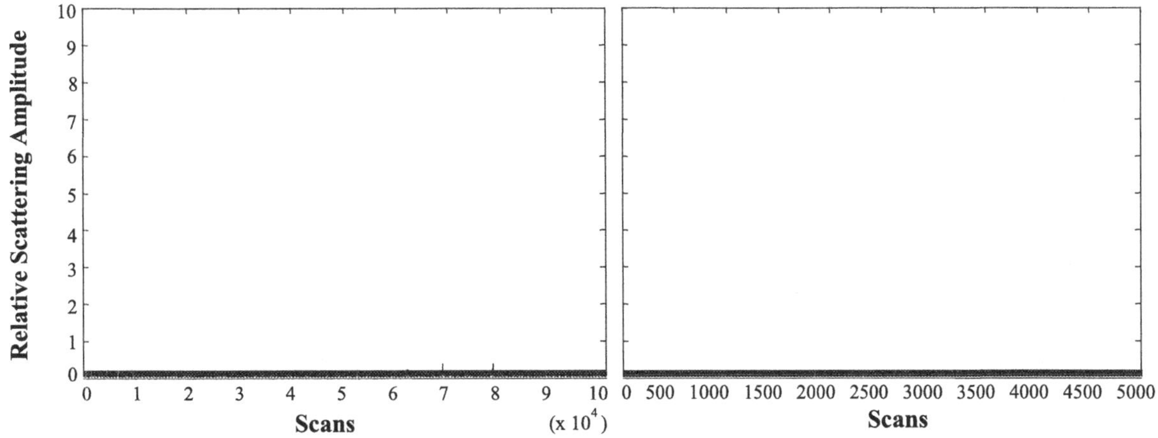

Figure 3.38 Gen-3 MALS measurements of purified laboratory water (LW). The particle count for this sample was 2/mL. The left-hand figure shows the MALS raw data output for 100,000 scans, or 8 seconds, of data. The right-hand figure shows the same data, for the first 5,000 scans (0.4 seconds).

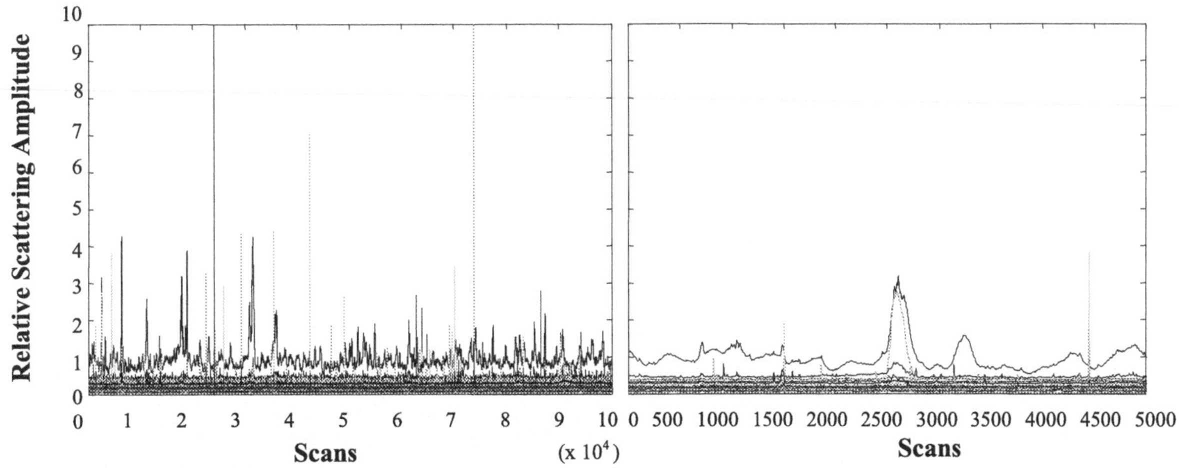

Figure 3.39 Gen-3 MALS measurements of Weymouth Plant Effluent (WPE). The mean particle count for this sample (based on five repeated measurements) was 108/mL. The left-hand figure shows the MALS raw data output for 100,000 scans, or 8 seconds, of data. The right-hand figure shows the same data, for the first 5,000 scans (0.4 seconds).

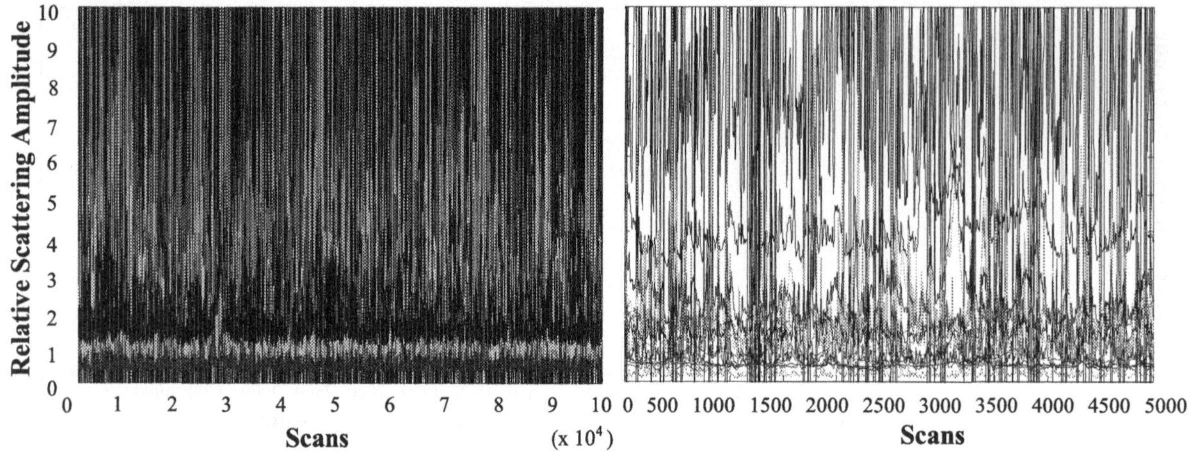

Figure 3.40 Gen-3 MALS measurements of raw Colorado River Water (CRW). The mean particle count for this sample (based on five repeated measurements) was 7709/mL. The left-hand figure shows the MALS raw data output for 100,000 scans, or 8 seconds, of data. The right-hand figure shows the same data, for the first 5,000 scans (0.4 seconds).

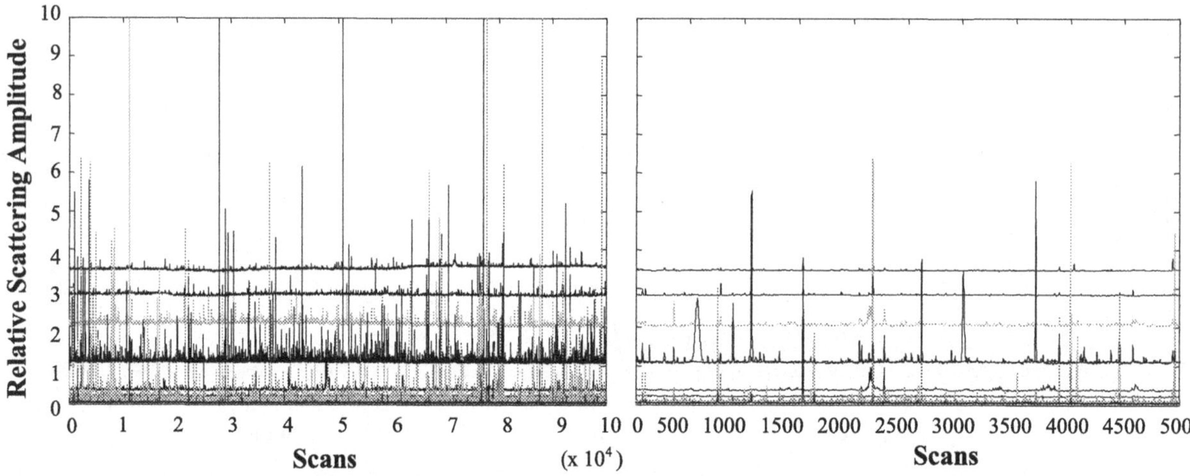

Figure 3.41 Gen-3 MALS measurements of raw Irvine Ranch Ground Water (IRGW). The mean particle count for this sample (based on five repeated measurements) was 615/mL. The left-hand figure shows the MALS raw data output for 100,000 scans, or 8 seconds, of data. The right-hand figure shows the same data, for the first 5,000 scans (0.4 seconds).

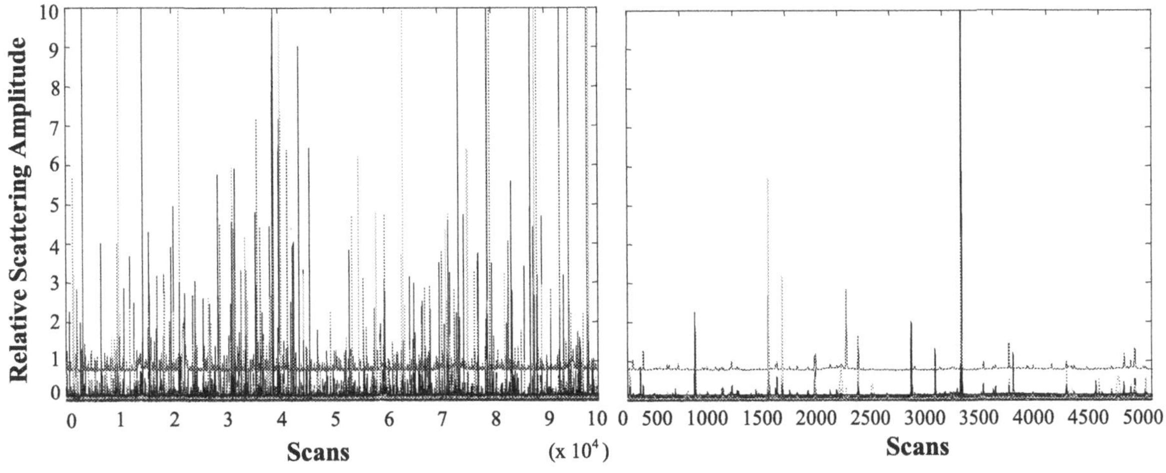

Figure 3.42 Gen-3 MALS measurements of raw Orange County Ground Water (OCGW). The mean particle count for this sample (based on five repeated measurements) was 532/mL. The left-hand figure shows the MALS raw data output for 100,000 scans, or 8 seconds, of data. The right-hand figure shows the same data, for the first 5,000 scans (0.4 seconds).

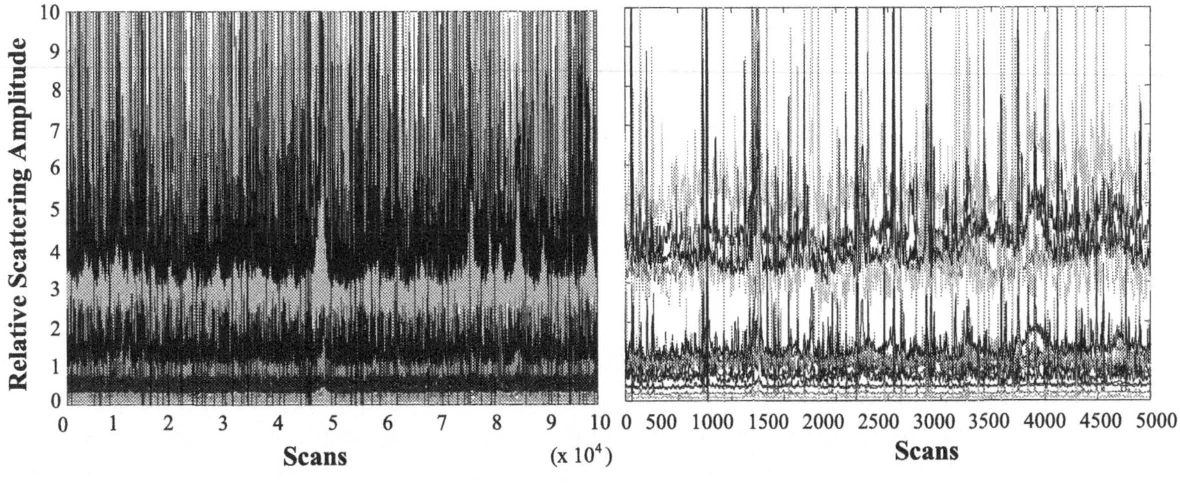

Figure 3.43 Gen-3 MALS measurements of raw California State Project Water (SPW). The mean particle count for this sample (based on five repeated measurements) was 6898/mL. The left-hand figure shows the MALS raw data output for 100,000 scans, or 8 seconds, of data. The right-hand figure shows the same data, for the first 5,000 scans (0.4 seconds).

pass through the laser beam. In each of the following figures, 16 channels of data are shown, with time as reflected in scan number (12,500 scans per second) on the x-axis and response in volts on the y-axis. The system saturated at 10 volts. All data will be shown using the same scale for comparison purposes. These data include: MALS measurements of the same water samples analyzed in Set 1 in Table 3.30 using the Gen-3 instrument; and MALS measurements of the following waters using the Gen-4 instrument: purified laboratory water (LW), "low" particulate count Weymouth plant effluent (WPE), "high" particulate count WPE, Weymouth plant influent (WPI) – a blend of Colorado River water and (California) State Project water, and Orange County Ground Water (OCGW).

For these six water samples, there was at least a qualitative correlation between the particle count results and the MALS raw data for these water matrices. The higher the particle count, the more "busy" was the MALS response.

The ability for MALS to detect target particles such as *Cryptosporidium* in a water matrix was dependent upon the occurrence of single particle events. Qualitative assessment of these six water matrices appeared, for the Gen-3 instrument, to provide for single particle events for LW, WPE, IRGW and OCGW. However, both the CRW and the SPW were exceedingly "busy" – particles were passing through the detection zone nearly continuously, so it is doubtful that identification could take place in these media with the Gen-3 configuration of laser and detectors.

Irvine Ranch ground water was not available at that time that the Gen-4 MALS instrument came on line. Examples of raw MALS data from the following waters are shown in the following figures.

- Purified laboratory water (LW);
- "Low" particle count Weymouth Plant Effluent (WPE);
- "High" particle count Weymouth Plant Effluent (WPE);
- A blend of Colorado River raw water (CRW) and State Water Project water (SWP), also called Weymouth Plant Influent (WPI);
- Orange County Ground Water (OCGW).

Figures 3.44 through 3.48 were representative samples typical of the types of water that were investigated during this project. These were measurements performed in the Gen-4 MALS instrument, but were not – except for the low particulate WPE sample – the same samples for which the water chemistry is reported in Table 3.30.

In order for MALS to operate in a particular water sample, there must have been periods of quiescence during which no matrix particles are passing through the patrol volume. If there were no such time, events were multiple events, and therefore, target particles could not be distinguished.

The LW data in Figure 3.44 show virtually no events, which was typical for purified laboratory water (LW). The two WPE samples showed the possible variation in the WPE samples, and the differences between these samples, which was reflected in event rate. There were many more events per time, and many more large events, in the high particulate WPE samples. The particle count for these two samples (> 2 µm) was 17.8 per mL and 132.4 per mL, which fell roughly between the 50th percentile and 90th percentile, respectively, for typical finished waters (McTigue et al., 1998).

The WPI, which was a blend of CRW and SPW, showed so many overlapping events that it was virtually impossible for the tested configuration of MALS to operate in this environment. This did preclude MALS from working in raw waters, however, but the matrix water may have

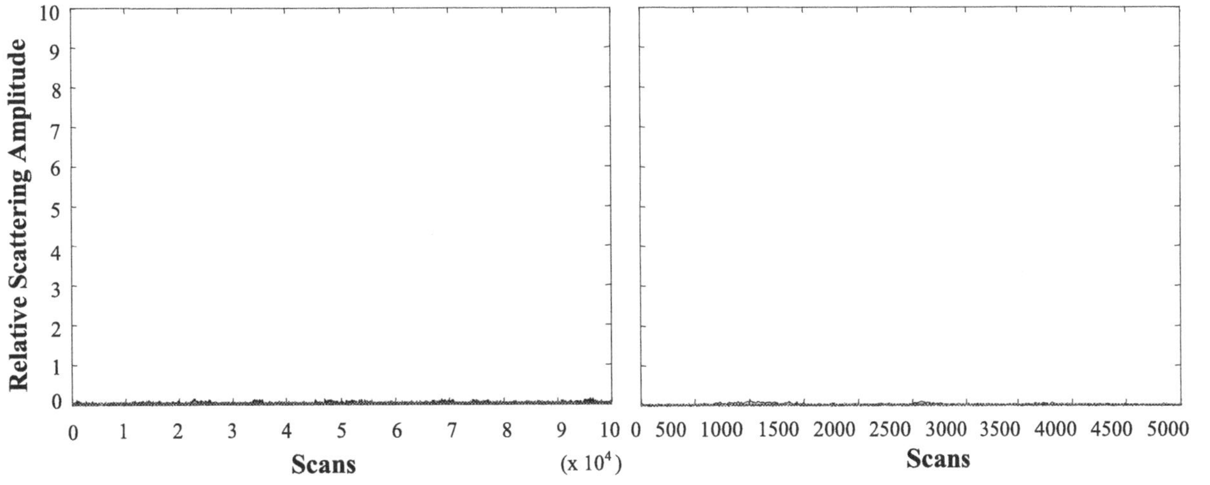

Figure 3.44 MALS measurements of typical purified Lab Water (LW) using the Gen-4 MALS instrument. The left-hand figure shows the MALS raw data output for 100,000 scans, or 8 seconds, of data. The right-hand figure shows an expanded view of the same the same data, for the first 5,000 scans (0.4 seconds).

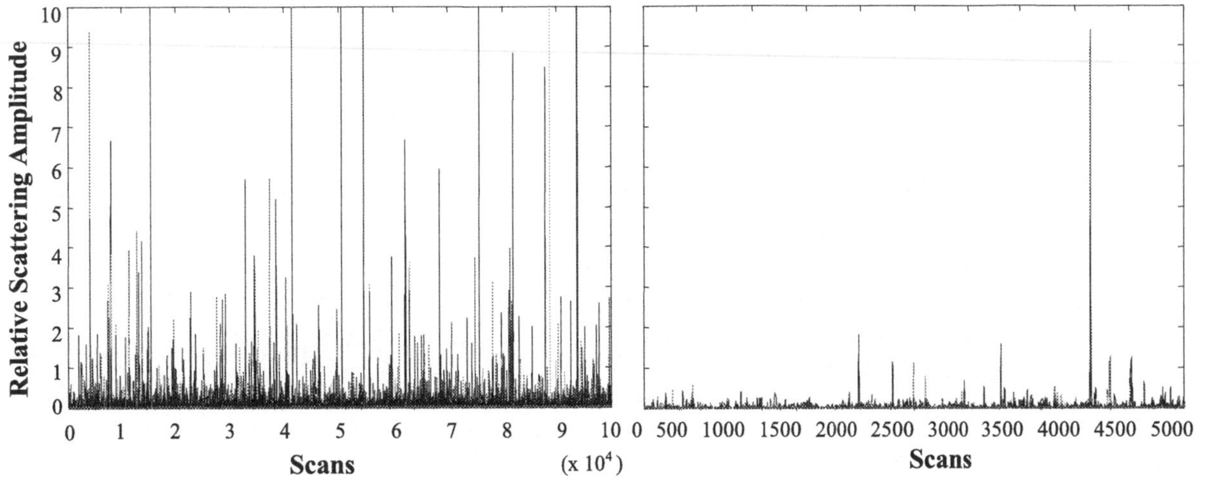

Figure 3.45 Gen-4 MALS measurements of an example of low particulate WPE. The particle count (> 2 μm) for this sample was 17.8 per mL. The left-hand figure shows the MALS raw data output for 100,000 scans, or 8 seconds, of data. The right-hand figure shows the same data, for the first 5,000 scans (0.4 seconds).

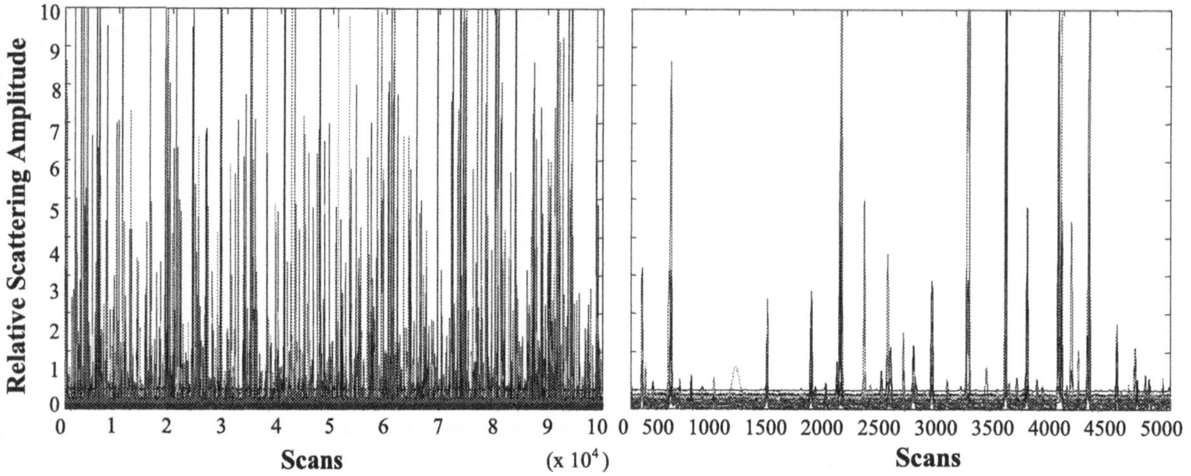

Figure 3.46 Gen-4 MALS measurements of an example of high particulate WPE. The particle count (> 2 μm) for this sample was 132.4 per mL. The left-hand figure shows the MALS raw data output for 100,000 scans, or 8 seconds, of data. The right-hand figure shows the same data, for the first 5,000 scans (0.4 seconds).

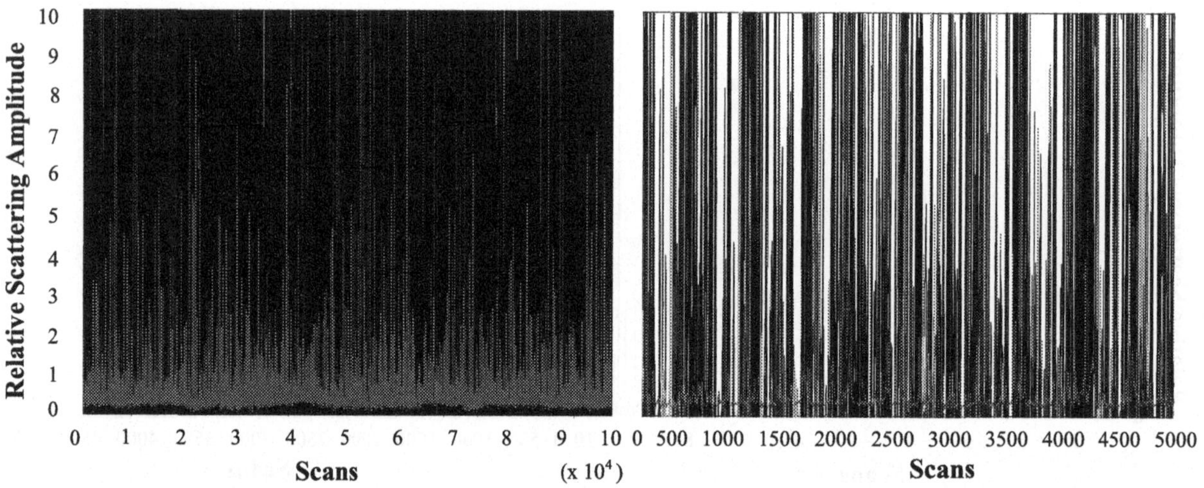

Figure 3.47 Gen-4 MALS measurements of Weymouth Plant Influent (WPI), a blend of raw Colorado River water and raw State Project water. The particle count (> 2 μm) for this sample was 6458.0 per mL. The left-hand figure shows the MALS raw data output for 100,000 scans, or 8 seconds, of data. The right-hand figure shows the same data, for the first 5,000 scans (0.4 seconds).

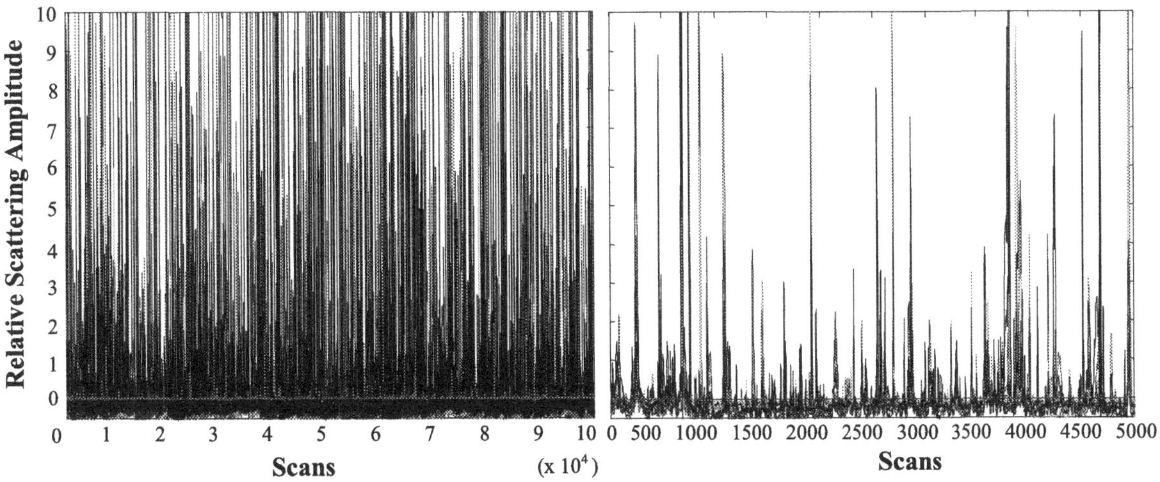

Figure 3.48 Gen-4 measurements of a sample of Orange County Ground Water (OCGW). The particle count (> 2 μm) for this sample was 403.1 ± 6.4 per mL. The left-hand figure shows the MALS raw data output for 100,000 scans, or 8 seconds, of data. The right-hand figure shows the same data, for the first 5,000 scans (0.4 seconds).

required pre-filtering using a large filter size (say 50 μm pore size) to collect the large particles, or the instrument could utilize a narrower laser beam to minimize multiple events.

The particulate content of the OCGW sample fell between the WPE samples and the WPI sample, as was expected by the particle count (17.9 per mL and 132.5 per mL for WPEs, 403.1 per mL for OCGW, and 6458.0 per mL for WPI). While the fraction of quiescent time for the OCGW was lower than the WPE, it appeared that MALS may operate in a ground water environment without any pre-filtering or optical modifications.

Since nearly all of the measurements were performed in WPE, typical matrix water used is shown by the WPE samples in Figures 3.45 and 3.46.

CHAPTER 4
SUMMARY AND MAJOR FINDINGS

The overall goal of this project was to perform a proof-of-concept to evaluate the sensitivity, viability and accuracy of the multi-angle light scattering (MALS) technique to detect and identify *Cryptosporidium* oocysts under various matrix conditions.

While the MALS technique may potentially be used for a broad set of waterborne pathogens, *Cryptosporidium* was chosen as the microorganism of interest for the following reasons:

- It is resistant to standard treatment techniques;
- It is not easily culturable and thus detection and identification typically takes longer and is more expensive than other pathogens such as *E. coli*;
- Since the 1993 Milwaukee incident there has been a high awareness and interest in *Cryptosporidium*;
- It is more physically stable and morphologically uniform than vegetative bacteria in environmental or treated waters, so the results here represent a higher performance range for MALS relative to vegetative microorganisms.

An increasingly challenging set of water types, from purified laboratory water to raw source waters were tested to determine the ability of MALS to operate under various water conditions. In addition, the MALS technique was tested to determine if MALS could provide sufficient sensitivity to detect morphological changes of the *C. parvum* oocyst (if any) brought about by natural causes such as excystation or through disinfection such as ozonation or ultraviolet (UV) treatment.

Two unforeseen factors affected preliminary testing: (1) Early MALS instrumentation (Gen-3) was sensitive to cleaning and disinfection, frustrating stable and consistent measurements, and (2) the initial oocyst samples provided were not of sufficient optical quality in order to generate good optical fingerprints. These factors were corrected respectively by: (1) modifying the MALS design in order to remove the detectors from direct contact with the water and disinfection solutions (as reflected in the Gen-4 instrumentation); and (2) performing a final flow cytometry cell sorting step on the oocyst samples to remove any extraneous particles such as bacteria and organic detritus. These steps significantly improved the data quality and provided a stable measurement platform that enabled comparative experiments to be performed.

The first set of tests of MALS performance involved the testing of various water types from highly purified laboratory water (very low particulate concentrations) to raw source waters from the State Water Project and the Colorado River Water (very high particulate concentrations).

MALS operates by measuring a single particle per time as it passes through the laser beam. In order to minimize the occurrence of more than one particle in the beam at a time, which confounds analysis, the laser beam width can be modified to operate in a specific particle concentration environment: narrower for higher particulate content, and wider for lower particulate content. The tradeoff for operating in higher particulate water is lower volume measured per time (Effective Patrol Volume Rate – EPVR). For example, a narrower beam can operate in higher particulate waters but takes longer to measure a fixed volume than a wider beam that requires lower particulate water. The laser beam width used for this project was typically about 1.5 mm in width, generating a Patrol Volume of approximately 225 nanoliters. It was found that this particular design worked well for waters with particle counts (of size greater than 2 μm) less than

1,000 particles/mL. Therefore, for the current effort, operation in raw surface waters (SPW, CRW and WPI) was found to be not practical. However, it should be noted that this not a fundamental limitation of the MALS method, but rather a choice of design.

Initial measurements using the system in treated effluent waters (in this case, Metropolitan Weymouth Plant Effluent, or WPE) and raw ground waters (from Orange County) indicated that the Interference Factor (a measure of the loss due to the matrix interference, defined as the ratio of identifications per volume for identical spikes in matrix water versus purified near-zero particulate laboratory water) was unacceptably high – effectively 100% for the ground water and more than 80% for the treated effluent water.

A series of experiments were performed to investigate the cause of the high Interference Factors. These experiments revealed that the high Interference Factors were not:

- An experimental anomaly – they were reproduced with repeated measurements;
- Due to experimental procedures caused, for example, by sticking of the oocysts to tubing or test vessels;
- Caused by losses due to multiple particle events caused by high concentrations of oocysts;
- Due to the chemical differences between purified laboratory water, in which the system is trained to recognize the oocysts, and the matrix water.

Modifications were made to the MALS system to improve the amplifier electronics and the feature extraction software. Based on these modifications, the performance of the system improved dramatically. Interference Factors improved to 10% to 50%, the value of which was dependent upon the data analysis parameters. This represented a significant breakthrough and demonstrated the practical ability to detect and identify *Cryptosporidium* in drinking water using a real-time continuous system.

Identification Rates (the ratio of *Cryptosporidium* oocysts identified correctly to the total number of oocysts detected passing through the system) of 11% to 45% were achieved, with accompanying False Positive Rates (the ratio of matrix water particles incorrectly identified as *Cryptosporidium* to the total number of matrix particles passing through the system) of 0.3% to 3%. The variability in Identification Rates and False Positive rates was found to be controllable though the analysis software. Much like receiver-operator characteristic (ROC) curves typically used for detection analysis (c.f. Macmillan and Creelman, 1991), a higher Identification Rate can be achieved at the cost of a higher False Positive Rate, and conversely a lower False Positive Rate can be achieved at the cost of a lower Identification Rate. The MALS system can be "tuned" via software to provide either case, or something in the middle, for a potential user of the system, based on needs and requirements.

Experiments were performed to determine the estimated limit of detection (ELOD) for two water matrices: purified laboratory water and finished drinking water (WPE). In both cases, the ELOD was determined through a serial dilution experiment. In the purified laboratory case, individual spiking experiments were performed at low concentrations to verify the serial dilution experiments. Because the MALS system is a flow-through system, the ELOD was a function of time. Volume and time are related through the speed of the water passing through the system. Results for the ELOD for laboratory water, a sample of finished drinking water, and the theoretical best possible performance (based on ideal conditions that were not attained in this effort) for a single read-head, in oocysts/mL are shown in Table 4.1. The theoretical best performance shown

Table 4.1
Summary of Estimated Limits of Detection (ELOD) in oocysts/mL for purified laboratory water, finished matrix water, and the theoretical best performance of a single read-head

	Detection time				
Matrix	1 min	10 min	1 hr	5 hrs	10 hrs
Purified Laboratory Water[1]	6.7	0.7	0.1	0.02	0.01
Purified Laboratory Water[2]	10	0.1	0.004		
Finished Drinking Water	77.0	7.7	1.3	0.3	0.1
Theoretical Best, Signal Head	0.05	0.006	0.009	0.0002	0.0001

Notes:
1. Results generated using a single serial dilution experiment.
2. Results generated from a group of spikes at specific concentrations.

in this table was independent of matrix water. Estimates for the single spike results are not shown beyond one hour because of the limited data available.

The actual concentrations of oocysts that occur during contamination outbreaks (e.g. Milwaukee 1993) are not known – however, oocyst concentrations are likely to be much higher than levels typically found during national occurrence surveys (for example, the EPA's ICR data collected from July 1997 to December 1998- Reference: ICR; USEPA, 1996), perhaps by as much as one to several orders of magnitude. Relatively small changes to the system could improve the limit of detection of the MALS system reported in Table 4.1. An example would be to increase the flow of water through the read-head by a factor of five. This speed (which has been tested independently but is not reported here) would drop the limit of detection to several hundred oocysts per liter in a reasonable time (for example, one hour) in finished drinking water. These levels may be of interest to use MALS as an early warning system for cases of contamination events that show short-duration, high concentrations of oocysts.

Based on improvements already achieved, it is possible that improved feature extraction, detection schemes and faster electronics could improve the limit of detection by another order of magnitude. Theoretical best performance – which is unlikely to be fully met but may be approached with improved instrumentation – can nearly reach limits of detection that would enable the system to measure oocysts approaching levels of a few oocysts per liter. At this level of measurement, the MALS system could be used for compliance verification. If a single read-head could not reach these levels in a reasonable amount of time (say one hour), read-heads could be ganged together to get the benefit of a linear decrease in the detection time for a given detection limit.

It was found that MALS was able to distinguish between live infectious oocysts from those treated by heat and ozone. It is apparent from these results that heat treatment and ozone treatment generate morphological changes significant enough in the oocysts to be detectable by MALS. In the case of heat treatment, the internal sporozoites are damaged or destroyed, and in the case of ozone treatment, there was likely significant surface and morphological damage to the oocysts.

In addition, MALS was able to discriminate excysted oocysts from non-excysted oocysts. This is in contrast with protein-based methods such as IFA or DNA-based methods such as PCR. However it was also found that MALS cannot discriminate between ultraviolet (UV) treated oocysts and infectious untreated oocysts based on the experiments performed in this project. This is not an unexpected result – UV inactivation takes place on the molecular level and there is no immediate morphological change in the oocysts, therefore no morphological change can be detected.

In summary, it was shown that MALS can distinguish between *Cryptosporidium* oocysts and background finished water matrix particles. This ability translated into useful limits of detection, acceptable levels of interference with an example of a finished drinking water matrix, and relatively low false positive rates. The ability to detect *Cryptosporidium* oocysts at a concentration of a few hundred to a thousand oocysts per liter in reasonable time may suggest, upon further development and testing, that MALS could provide an early warning, rapid response tool in treatment plants and distribution systems to detect large scale outbreaks, either naturally caused by storm events, etc. or induced purposefully with the intent to cause harm or panic.

Evolutionary improvements in the development plan executed during this effort included progressive instrumentation improvements that have not exhausted the scientific potential of MALS. New technologies (e.g., optics, detectors, electronics) can be applied to the physical scattering process to deliver improved results, including for example, operation in more challenging waters like source or raw water in the future and improved performance parameters: higher Identification Rates, lower False Positive Rates, and lower Limits of Detection for a given time.

Since the terrorist attacks on the United States on September 11, 2001, there is a new environment of enhanced security and higher awareness that would be better served with monitoring systems that can not only detect large scale rapid changes, but can also minimize public panic by being able to rapidly isolate and respond to purposeful contamination. This can only be done with a real-time, continuous on-line remote and autonomous system. The MALS technology shows promise in being able to provide this type of monitoring, in the present case, for a common and dangerous pathogen, *Cryptosporidium parvum.*

CHAPTER 5
RECOMMENDATIONS AND AREAS FOR FUTURE RESEARCH

One of the limitations of the results presented in this report is the limited number and extent of the measurements, for example the results that were used to generate an estimated limit of detection (ELOD). Repeated and more detailed experiments should be performed in order to generate more complete statistics and gain better quantification of the sensitivity and limit of detection of the MALS system. This is particularly true for the lower spiking concentrations, which are intrinsically difficult experiments to perform.

While the current research demonstrated that the MALS technology can be used to detect and identify *Cryptosporidium parvum* oocysts in purified laboratory and finished drinking water, the result is fundamentally limited due to the small number of water samples tested. The MALS signature of *Cryptosporidium* oocysts was found to be different from the matrix water particles in the limited number of matrix water samples tested, but a significantly broader set of treated water sample types from various geographic localities, treatment plant types and source waters should be explored to broaden the potential application of MALS for the water industry. If the first and most likely application of the MALS system is for post-treatment water, both in the plant and in the distribution system, the performance of the system should be measured against a wide and representative sample of types of post-treatment waters expected.

The proof-of-concept work focused on the ability of MALS to detect and identify *C. parvum* in treated drinking water. While *Cryptosporidium* is an organism that is of particular interest because of its resistance to standard treatment techniques, there are a number of other waterborne pathogens that are of concern to the water industry in the treatment and distribution system, including *E. coli* and other enteric bacteria, aerobic spore formers such as *B. subtilis*, other protozoa like *Giardia lamblia* and various algae of interest to the water industry such as *Chlorella*. The MALS system should be tested and challenged with these organisms to determine the ability for the system to speciate and distinguish between them.

Results from this research indicated that a single MALS instrument could not simultaneously operate optimally (maximizing Patrol Volume Rate) in both low particulate water (finished, treated, and/or filtered) water and source waters. Therefore, this project concentrated on low particulate waters as a first and nearest term application of real-time, continuous, on-line monitoring. It would be of interest to the water industry to test the capabilities of the MALS system in a raw or source water environment. The ability to place MALS instruments on the influent and effluent of a treatment plant, for example, might provide a means to optimize treatment regarding microorganisms. Such a real-time, on-line, continuous system might be able to minimize disinfectant by-products while monitoring for and minimizing microbiological risk.

The experimental development plan included evolutionary instrumentation improvements that have not exhausted the scientific potential of the technique. New improvements, some of them as simple as running the water faster through the system, can be applied to the physical scattering process to deliver improved results in the future. It is not unreasonable to expect a 10 – 100 times improvement in performance as the system improves. Examples of these improvements are: (a) running the water faster through the system, with a potential improvement of 2 to 5 times; (b) improved feature extraction techniques, with a potential improvement of 2 to 3 times; (c) improved electronics with a potential improvement of 2 to 4 times; and (d) improved optics with a potential of 2 to 3 times. Revisiting the types of experiments performed in this project would be useful once these new improvements have been achieved, in order to measure and quantify the improved performance of the MALS technology.

In the current environment of high terrorist risk, and with the potential vulnerability of US water systems to low technology terror attack, a pilot project with several to tens of instruments placed in the distribution system would provide a real-world and meaningful test of the technology. The only practical and cost-effective method to prevent purposefully contaminated water from being consumed is to provide early warning via wide-scale, low cost, real-time, continuous, remote and autonomous monitoring. Such a system would provide a positive method to detect unusual microbiological (or other particulate-based) events and also lower the risk of panic from false alarms or false threats on water systems.

During this project, it was found that the existing sources for *Cryptosporidium* are limited, the sources for *Giardia* are unreliable, and the sample quality, while sufficient for standard batch method testing, was not sufficient for optical techniques such as MALS. In this new environment of a potentially constant or perhaps increasing threat of biological attacks on the US water infrastructure, there may be as many as hundreds of private and public groups looking for standardized samples that can provide consistent and reliable results, minimizing the uncertainty and shortage of quality microbiological samples.

REFERENCES

APHA, AWWA, and WEF (American Public Health Association, American Water Works Association, and Water Environment Federation). 1998. *Standard Methods for the Examination of Water and Wastewater.* 20th ed. Washington, D.C.: APHA.

Baribeau, H., P.A. Rochelle, R. De Leon, M.H. Stewart. 1999. Rapid Screening Assay for the Detection of *Cryptosporidium parvum* and Enteroviruses. Presented at AWWA Water Quality Technology Conference, Tampa, Fla.

Bohren, C.F., and D.R. Huffman. 1983. *Absorption and Scattering of Light by Small Particles.* New York: John Wiley & Sons.

Clancy, J.L., Z. Bukhari, R.M. McCuin, T.M. Hargy, C.R. Fricker, Z. Matheson, and N. Sikes. 2000. *New Approaches for Isolation of* Cryptosporidium *and* Giardia. Denver, Colo.: AwwaRF and AWWA.

Clancy, J.L., and J. Hansen. 1999. Uses of Protozoan Monitoring Data. *Journal AWWA*, 91(5):51–65.

Cooley, W.W., and P.R. Lohnes. 1971. *Multivariate Data Analysis.* New York: John Wiley & Sons.

Di Giovanni, G.D., J.W. Czajka, M.E. Schaffer, G. Tice, M.W. LeChevallier, and M. Abbaszadegan. 1999. A Fluorescent Homogeneous Format PCR Kit for the Detection of *Cryptosporidium parvum.* Presented at AWWA Water Quality Technology Conference, Tampa, Fla.

Duke Scientific. 1997. Certificate of Calibration and Traceability. Palo Alto, Calif.

Fitzgerald, W.F., and J. Rooney. 1996. *Numerical Bayesian Methods Applied to Signal Processing.* Springer-Verlag.

Goldstein, S.T., A.W. Hightower, and A.J. Bryant. 1996. Cryptosporidiosis: An Outbreak Associated With Drinking Water Despite State-of-the-Art Water Treatment. *Annals of Internal Medicine*, 124(5):459–468.

Knott, E.F., J.F. Shaeffer, and M.T. Tuley. 1985. *Radar Cross Section.* Norwood, Mass.: Artech House.

Kozwich, D., K. Johansen, C. Roehl, K. Landau, P. Roehl, and H.S. Woronoff. 1999. An Evaluation of a Novel Molecular Method for the Detection of *Cryptosporidium parvum* in Finished, Source and Ground Water Samples. Presented at AWWA Water Quality Technology Conference, Tampa, Fla.

Kramer, M.H., B.L. Herwaldt, G.F. Craun, R. Calderon, and D.D. Juranek. 1996. Surveillance for Waterborne Disease Outbreaks–United States, 1993–1994. *Morbidity & Mortality Weekly Report*, 45:1–15.

LeChevallier, M.W., and W.D. Norton. 1995. *Giardia* and *Cryptosporidium* in Raw and Finished Drinking Water. *Journal AWWA*, 87(9):54–68.

LeChevallier, M.W., W.D. Norton, R.G. Lee, and J.B. Rose. 1991. Giardia *and* Cryptosporidium *in Water Supplies.* Denver, Colo.: AwwaRF and AWWA.

Linden, K.G., and A.A. Mofidi. 1999. Measurement of UV Irradiance: Tools and Considerations. In *Proceedings of the AWWA Water Quality Technology Conference.* Denver, Colo.: AWWA.

MacKenzie, W.R., N.J. Hoxie, M.E. Proctor, M.S. Gradus, K.A. Blair, D.E. Peterson, J.J. Kazmierczak, D.G. Addiss, K.R. Fox, J.B. Rose, and J.P Davis. 1994. A Massive Outbreak in Milwaukee of *Cryptosporidium* Infection Transmitted Through the Public Water Supply. *New England Journal of Medicine*, 331:161–167.

Macmillan, N.A., and C.D. Creelman. 1991. *Detection Theory: A User's Guide.* Cambridge, England: Cambridge University Press.

McTigue, N.E., M. LeChevallier, H. Arora, and J. Clancy. 1998. *National Assessment of Particle Removal by Filtration.* Denver, Colo.: AwwaRF and AWWA.

Mofidi, A.A., H. Baribeau, P.A. Rochelle, R. De Leon, B.M. Coffey, and J.F. Green. 2001. Disinfection of *Cryptosporidium parvum* With Polychromatic UV Light. *Journal AWWA,* 93(6):95–109.

Nieminski, E.C., W.D. Bellamy, and L.R. Moss. 2000. Using Surrogates to Improve Plant Performance. *Journal AWWA*, 92(3):67.

PointSource Technologies. 2002. Internal documentation.

Quist, G.M. 2000. Real-Time On-Line Monitoring for Protozoa in Drinking Water. In *Proc. of the AWWA Water Quality Technology Conference.* Denver, Colo.: AWWA.

Quist, G.M., and H.E.G. Ix. 2003. Identification of Particles in Fluid. U.S. Patent 5,619,033.

Quist, G.M., and P.J. Wyatt. 1985. Empirical Solution to the Inverse-Scattering Problem by the Optical Strip Map Technique. *Journal of the Optical Society of America,* (Nov.):1979–1985.

Rochelle, P.A., D.M. Ferguson, T.J. Handojo, R. De Leon, M.H. Stewart, and R.L. Wolfe. 1997. An Assay Combining Cell Culture With Reverse Transcriptase-PCR to Detect and Determine Infectivity of Waterborne *Cryptosporidium parvum. Applied & Environment Microbiology,* 63:2029–2037.

Rochelle, P.A., M.M. Marshall, J.R. Mead, A.M. Johnson, D.G. Korich, J.S. Rosen, and R. De Leon. 2002. Comparison of *In Vitro* Cell Culture and a Mouse Assay for Measuring Infectivity of *Cryptosporidium parvum. Applied & Environment Microbiology,* 68:3809–3817.

Rose, J.B., C.P. Gerba, and W. Jakubowski. 1991. Survey of Potable Water Supplies for *Cryptosporidium* and *Giardia. Environmental Science & Technology,* 25:1393–1400.

Sachweh, D.A., W.D. Dick, and P.H. McMurry. 1995. Distinguishing Between Spherical and Nonspherical Particles by Measuring the Variability in Azimuthal Light Scattering. *Aerosol Science and Technology,* 23:373–391.

USEPA (U.S. Environmental Protection Agency). 1996. National Primary Drinking Water Regulation: Monitoring Requirements for Public Drinking Water Supplies: *Cryptosporidium, Giardia,* Viruses, Disinfection By-Products, Water Treatment Plant Data and Other Information Requirements. Final Rule. *Fed. Reg.*, 61:94:24354–24388 (May 14).

USEPA (U.S. Environmental Protection Agency). 2001. Method 1623: *Cryptosporidium* and *Giardia* in Water by Filtration/IMS/IFA. Office of Water, EPA-821-R-99-006, April 2001. Washington, D.C.: USEPA.

WHO (World Health Organization). 1998. *World Health Report 1998, Life in the 21st Century—A Vision for All.* Geneva, Switzerland: WHO.

Wyatt, P.J. 1968. Differential Light Scattering: A Physical Method for Identifying Living Bacterial Cells. *Applied Optics*, 7:1879–1896.

Wyatt, P.J. 1975. *Atlas of Light Scattering Curves.* Santa Barbara, Calif.: Science Spectrum.

Wyatt, P.J. 1989. Rapid Detection of Toxicants in Potable Drinking Waters by Laser Light Scattering. Final Report. Ft. Detrick, Md.: U.S. Army Medical Research and Development Command.

Wyatt, P.J., and C. Jackson. 1989. Discrimination of Phytoplankton Via Light-Scattering Properties. *Limnology and Oceanography,* 34(1):96–112.

Wyatt, P.J., K.L. Schehrer, S.D. Phillips, C. Jackson, Y.J. Chang, R.G. Parker, D.T. Phillips, and J.R. Bottinger. 1988. Aerosol Particle Analyzer. *Applied Optics*, 27:217.

TERMS AND ABBREVIATIONS

A/D board	analog-to-digital computer board used to transform analog electrical signals into digital signals
AwwaRF	Awwa Research Foundation
B. subtilis	*Bacillus subtilis*
C. parvum	*Cryptosporidium parvum*
CPU	central processing unit
CRW	Colorado River water, untreated water from the MWD Colorado River source
CsCl	cesium chloride
DAWN-B	the dual angle weighted nephelometer, batch system, a Wyatt Technology (Santa Barbara, Calif.) product
DI	deionized
DNA	deoxyribonucleic acid
DNM	did not measure
DPD	N,N-diethyl-p-phenylenediamine
E. coli	*Escherichia coli*
ELOD	Estimated limit of detection. Generally measured in particles/volume (for example, oocysts/mL). For this project, ELOD was an estimate of the effective lower limit of detection of concentration of target particles that can be measured by the MALS system.
EPA	(United States) Environmental Protection Agency
EPVR	Effective Patrol Volume Rate. Generally measured in mL/hr. The coincidence between the laser beam and the OSM optics generated a region that can create events if a particle of sufficient size passes through that region. This volume was called the Patrol Volume. Since an event will not occur unless a sufficient number of detectors receive a sufficiently large signal, the "Effective Patrol Volume" will be a function of power and direction of the scattered light, which in turn is a function of size, shape, and refractive index of a particle. The Effective Patrol Volume rate was the product of the water speed and the cross sectional area of the Effective Patrol Volume. In general, smaller particles will have a lower EPVR than larger particles of the same shape.
Event	When a particle passes through the laser beam of the MALS system, and the optical signal generated by this particle was detected coincidentally and at a high enough amplitude in several detectors, and "event" has occurred.
EVM	Effective Volume Measured, typically measured in mL, by the MALS system during an experiment (see EPVR also). The EVM was the product of the EPVR and the time of measurement. The EVM depends upon the target particle and the water speed through the system.

FCCS	flow cytometry cell sorting
FNR	False Negative Rate. Measured in percentage. The FNR was the ratio of the number of target particles that are missed by the MALS system to the total number of targets detected as passed through the system, for example, the ratio of the number of *Cryptosporidium* oocysts that were missed by the system – called "unknown" – to the total number of oocysts detected as passing through the system. The False Negative Rate (FNR) and the Identification Rate (IDR) were related by: IDR = 1 – FNR. A perfect FNR would be 0%.
FPR	False Positive Rate. Measured in percentage. The FPR is the ratio of matrix particles incorrectly identified as target particles (e.g., *Cryptosporidium*) to the number of matrix particle events. The False Positive Rate (FPR) is measured using a blank (unspiked) matrix sample. A perfect FPR would be 0%.
FWPE	Filtered WPE, produced in order to test the proposition that the water chemistry had an effect on the Interference Factor for *Cryptosporidium*. FWPE was produced by filtering WPE once through a 0.1 μm filter. FWPE typically had particle counts (for size > 2 μm) of < 10 particles/mL. The single pass filtration reduced particles by about a factor of 10.
Gen-1 MALS	An early version of the MALS technology, using batch samples contained in 20 mL glass vials. Real-time, but not continuous or online.
Gen-2 MALS	The second version of the MALS technology, wherein all of the optical systems were completely submerged. This system was used for early proof-of-concept testing. Real-time and continuous, but not appropriate for on-line usage.
Gen-3 MALS	The third version of the MALS technology, wherein the laser was taken out of direct contact with the water. The first flow-through system. While real-time, continuous, and appropriate for on-line usage, this system had serious maintenance flaws, diminishing it's effective use as a real system.
Gen-4 MALS	The final version of the MALS system tested in this project. All of the optics, electronics, and detectors were taken out of direct contact with the water. The Gen-4 MALS system was an appropriate proof-of-concept device.
HBSS	Hanks balanced salt solution
hr	hour
H_2O_2	hydrogen peroxide
Hz	hertz, or cycles per second
ICR	Information Collection Rule
ID_{50}	infectious dose for 50% of the population
IDR	Identification Rate. Measured in percentage. The IDR for this project was the ratio of correctly identified targets to the number of targets passing through the system, for example, the number of *Cryptosporidium* oocysts to the number of oocysts detected as passing through the system. The Identification Rate (IDR) was measured in purified laboratory water using a spike of known concentration. A perfect IDR would be 100%.

IF	Interference Factor. Measured in percentage. The Interference Factor was a measure of the decrease of the effective Identification Rate (or increase of False Negative Rate) in a real water matrix compared to an ideal purified laboratory matrix, i.e., it was a measure of the interference caused by the real water matrix.
IMS	immunomagnetic separation
IRGW	Irvine Ranch Ground Water
ISAR	inverse synthetic aperture radar
L	liter
LW	Purified Laboratory Water, very low particulate water (typically less than 3 particles/mL of size > 2 µm) used for measurements of target particles, or controlled experiments. LW was generated in this project using single pass reverse osmosis and multiple pass multiple filters with pore sizes down to 0.05 µm.
M	molar
MALS	multi-angle light scattering. The technique studied in this project which relied upon analysis of the light scattered at various angles from individual particles in order to determine the type of particle that generated that light scattering pattern.
MALS System Configuration	Since the Gen-4 MALS system was an experimental system, there were a variety of ways in which the system could be set up and the data analyzed. The system configuration was continually changed during the project in order to test various configurations and improve system performance. Optics, detectors, laser beam, electronics, and analysis software could all be changed in order to determine the best possible configurations. Typically, results from tests performed within a given configuration were comparable, while results from tests performed on different configurations were not.
MANOVA	Multiple analysis of variance. A well-known mathematical technique used to discriminate different groups of objects that have multiple variables, for example, two groups of students, male and female, that have different ages, heights, weights, etc.
mg	milligram
min	minute
mJ	millijoule
mL	milliliter
MWD	Metropolitan Water District of Southern California
µm	micrometer
NA	not applicable
ND	not detectable
NIST	National Institute of Standards and Technology
nm	nanometer
NM	not measured
NTU	nephelometric turbidity unit
NWW	Northwest Water (Warrington, England)

OCGW — Orange County Ground Water, well water provided by the Orange County Water District

Optical Fingerprints — Three-dimensional maps (generally displayed in two dimensions because paper is flat) that show the relative positions of the target(s) and/or matrix(ces) after processing by MANOVA and other analytical processes. An Optical Fingerprint must have at least two targets/matrices, since it represented relative light scattering differences between two or more species. Optical Fingerprints were generated by making hundreds or thousands of measurements of individual particles under pristine scattering conditions. Once they are made, Optical Fingerprints were then used to make identification of unknown particles whose light scattering patterns were collected continuously and in real-time.

OSM — Optical sensor module. The integrated optics and detector used to: (a) limit the field of view of the detector in order to view only a portion of the laser beam; and (b) direct scattered light into a solid state detector that transformed the optical power scattered into electrical signals.

OSMAC — OSM array carrier. The physical piece used to align and hold the OSMs in order to place them at fixed angles to the scattered light.

Patrol Volume — the volume defined by the intersection of the laser beam and the combined optical fields-of-view of the OSMs

PC — particle count

PCPA — Poisson Change Point Analysis, a method used to evaluate the time series of target events (such as *Cryptosporidium*) produced by the MALS system. PCPA enables the optimal estimation of the time at which a change occurred and the mean values (e.g., concentration of particles) before and after the change.

PSL — polystyrene latex

PSL spheres — polystyrene latex spheres, standard and uniform light scattering standard spheres, used as MALS standards and system performance testing

RO — reverse osmosis

RT-PCR — reverse transcriptase-polymerase chain reaction

sec — second

SPW — State Project water, untreated water from the MWD California State Water Project source

Target Particles or Target — Particles or microorganisms that are of interest. Upon identification by the system, particles are either Targets (something of interest like a *Cryptosporidium* oocyst or a matrix particle) or they are unknown (a particle that does not fall into any of the classes of particles captured by the Optical Fingerprint).

TDS — total dissolved solids

TOC — total organic carbon

TSS — total suspended solids

t-test — the standard student's t-test, a well known statistical technique

UNK	unknown—a particle that was detected, but not identified by the MALS system
USEPA	United States Environmental Protection Agency
UV	ultraviolet
WPE	Weymouth Plant Effluent, product water from MWD's Weymouth Filtration Plant, in La Verne, Calif. WPE was the primary water used for drinking water tests.
WPI	Weymouth Plant Influent, input water to the MWD Weymouth Filtration Plant. WPI was a mixture of State Water Project Water and Colorado River Water.
WSLH	Wisconsin State Laboratory of Hygiene

IWA Publishing's authorised EU representative for General Product Safety Regulations is Diane D'Arras, 15 rue Duret, 75116 Paris, France, e-mail: safety@iwap.co.uk.

Printed and bound by CPI Group (UK) Ltd, Croydon, CR0 4YY
17/09/2026
02218811-0011